TREO MODEL OF
STRUCTURE AND
WORKING OF
UNIVERSE

Cosmic Code

DR. ASHOK SAXENA

INDIA · SINGAPORE · MALAYSIA

Notion Press

No.8, 3rd Cross Street, CIT Colony,
Mylapore, Chennai,
Tamil Nadu – 600004

First Published by Notion Press 2020
Copyright © Dr. Ashok Saxena 2020
All Rights Reserved.

ISBN 978-1-64828-887-6

For any query regarding this research work, please contact:
mail2ashoksaxena@gmail.com
91-9837153974

Dedication

"This work is dedicated to my late Parents,"

With sweet remembrance of late

MR. Y.K. Saxena

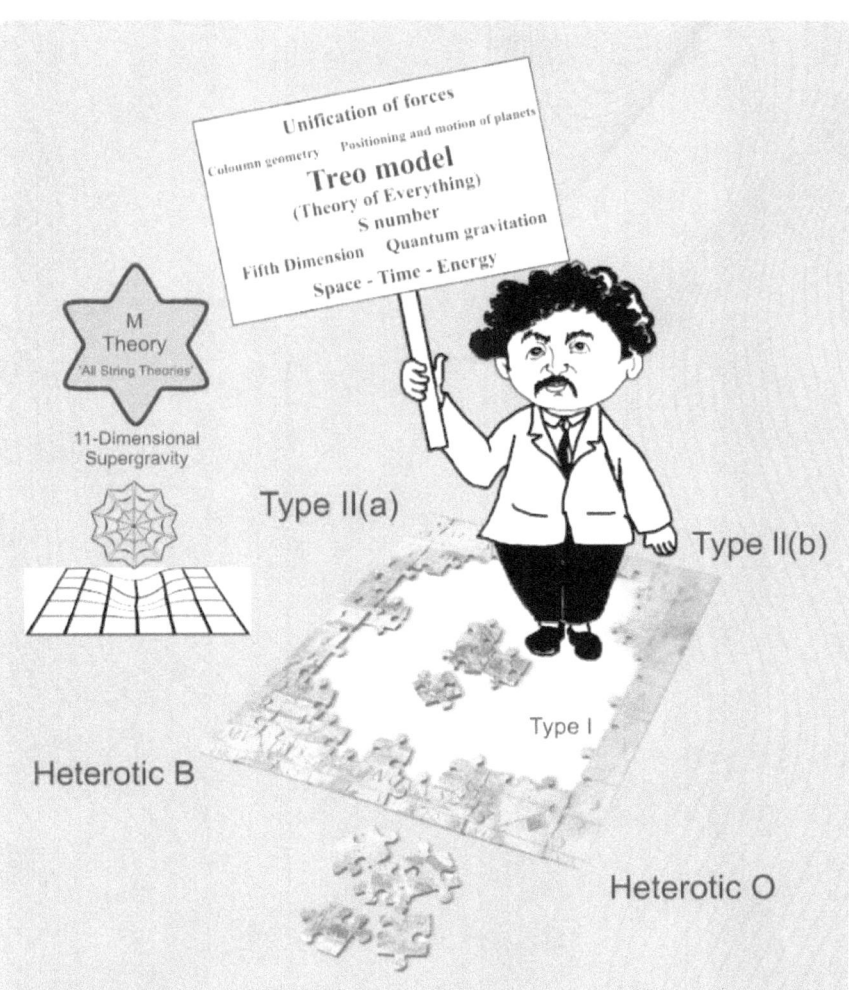

Dragon: as proposed by Dr Stephen Hawking

Uses Planck's units (God's units)

To describe working and properties of Space – matrix

(An Omnipresent, Omnipotent Reacting Force)

To unveil all mysteries of creation

(To decipher HIS script written every where)

Contents

Forwarded and Peer Reviewed By

1. Dr. R.C. Saxena

   M.Sc. (Physics), Ph.D. (London, UK)

   Senior Scientist, National Physical Laboratory, New Delhi.

2. Dr. H.P. Sinha

   M.Sc. Ph.D. (Physics),

   Director, Dau Dayal Institute of Vocational Education,

   Agra university, Agra

3. Mr. Shashi Kant Bhandari

   B.Tech, M.Tech (HONS)

   Astt. Professor, Department of Mechanical

   Engineering, Amity University, Greater Noida.

4. Dr. Anurag Saxena

   M.Sc. (Maths), Ph.D. (I.I.T. Kanpur)

   Indra Gandhi National Open University, New Delhi.

5. Dr. Avinash Mathur

   M.Sc. (Physics), Ph.D.

   Principal Scientist,

   Council of Scientific and Industrial Research,

   New Delhi.

Dr. R.C. Saxena

Senior Scientist, National Physical Laboratory, New Delhi.

Initially I encouraged Dr. Saxena to popularize his hypothesis of formation of an 'ECLIPTIC WINDOW ' during total solar eclipse observed in Agra, in the year 1995, and also helped him to present this work at various national and international forums.

Dr. Saxena started to build this treo model of universe about 25 years back, but initially I was very much suseptical with his new approach to substitute, Physical units in terms of Planck's units in place of SI units to get new derived values of universal constants?

He used to say that his proposed model is correct, only if this could provide explanation to all answered and get answers to all unanswered questions to project a clear and complete picture of universe. By this approach he could carve the 'theory of everything' which is explained in his four books.

IN THIS WORK; 'ENERGY' as one new fifth positive dimension of universe has been added. With this the long awaited Unification of all four basic forces has been achieved for the first time.

He could demonstare that how gravitational fields are formed by union of graviton coloumns; which explained true meaning of gravitation and revealed quantum gravitation. In this work, quantum physics (Physics of small) and general theory of relativity (Physics of big) has been successfully put in same frame.

He clarifies, that the composition of all matter is by integral number of Cmue (composite mass of unit electron) or mass units; and how fine structure constant calculates the composite mass energy of unit electron.

The model decides positioning criteria of all planets and satellites in solar system and the baby bodies of other stars. It predicts life span of our universe and mode of its death and its next rebirth.

I wish him success and good health.

DR. BHIM RAO AMBEDKAR UNIVERSITY AGRA

Dr. H.P. Sinha
Ex-Director, Dau Dayal Institute of Vocational Education,
Khandari, Agra-282002.
e-mail sinhahp@datainfosys.net

It is a matter of great pleasure to say a few words about this work?

Newton, Galileo, Kepler and others established the basic laws of physics and mechanics, followed by laws of thermodynamics which formed the basis of different heat engines. Towards the end of nineteenth century radio waves, X-rays, radioactivity, photoelectric effect were discovered.

In year 1900 Max Planck introduced the concept of photons & Planck's constant, elementary particles were discovered and theory of relativity and general relativity describing gravitational forces were proposed by Albert Einstein. Many scientists like Bohr, Feynman etc. developed physics to the present form.

Gravitational constant, Planck's constant, Speed of light, Planck's units, Mole theory all has got a fixed numerical value derived and verified experimentally from time to time.

The basic question before all the physicists was why these constants have a universal fixed value? Four different types of basic forces are observed in nature and physicists have been trying to find a unified theory for all these forces and to derive all universal constants, which should be true for macro and micro levels of universe.

Dr. Ashok Saxena made an attempt to solve this mystery. He has proposed a hypothesis of a basic string called treo and assumed that the entire universe, space, time, energy etc. consist of this basic unit. From this theory he has derived the expression for mass, space, time, energy etc. and the value of universal constants. On its basis he was able to explain the entire picture of cosmology and of universe, expanding nature of universe, presence of life in another galaxy or star system and several other unanswered questions.

He also explained the structure of atom, and basis of formation of all 118 elements as categorized in periodic table.

I hope the proposed Treo model of universe will be appreciated by various physicists, cosmologists and other scientists.

Shashi Kant Bhandari,

**Astt Professor, Department of Mech. Engineering
Amity University, Greater Noida. Ph 9810096233**

I feel great while writing few words for your work, specifically the latest one.

"The 20th Century had seen a lot of development in the field of science, contributed by so many scientists. Albert Einstein, Neil Bohr, Max Plank, Richard Feynman, Roger Penrose, Stephen Hawkins, to name a few. But there had been a challenge before all, How to get a Unified Theory of all different Natural Forces."

Gravitational Constant, Planck's Constant, Speed of Light, Mole Theory, All have got a fixed numerical value, derived at different times by different persons. As a science student, the question had been always striking to my mind that there has to be some unified theory, working equally at all levels macro & micro level of sub atomic parts.

Dr. Ashok Saxena had dared to solve this mystery. He had put before us a hypothesis, by which we can derive and get the numerical values of all the above natural constants, His hypothesis gives us the entire picture of Cosmology, answers about the age of the universe, its ever expanding nature, etc etc.

Thermodynamics is one of major subject of Physics & Mechanical Engineers. All heat Engines, power plant, Aviation Engines, Refrigeration systems are working on it. Entire Thermodynamics is based on three hypothesis given nearly 150 years back. Any hypothesis becomes the Law, when it does not find any contradiction, as in the case of thermodynamics named as I, II and III Law. Similarly Dr. Saxena's work can be treated as hypothesis for the time being and it will become a Natural Law, in the due course of time, in the absence of any contradiction.

Ignou (INDRA GANDHI NATIONAL OPEN. UNIVERSITY)

THE PEOPLE'S UNIVERSITY

DR. ANURAG SAXENA

Professor of managment

I have gone through the work of Dr. Ashok Saxena. He has simply replaced the SI units of length and time, (which are being used to get conventional values of universal constants), with the Planck's unit of least length and least time and thus he could re-calculated new values of universal constants.

These new values of universal constants calculated by these simple replacements put all constants togather, as they all seems to be governed by S number i.e. the value of Planck's frequency 1.85539×10^{43} (it is said to be COSMIC RHYTHM; at which all bound treos of omnipresent space matrix of universe are vibrating). New values, thus obtained makes them live and self explainatory, as each constant gets some purpose to explain universe beautifully.

These descriptions of universe are compiled in four books of Dr. Ashok Saxena as Treo model.

I wish him successes and good health.

Dr. AVINASH MATHUR

Principal Scientist,
Council of Scientific and Industrial Research,
New Delhi.

The single unified energy that runs the universe is the supreme lord, the cause of all causes.There is therefore likely to be a unified simple phenomenon that explains everything.

All string theories and M theories and all pure mathematical models postulates that whole universe is made up of one dimensional tiny Strings each of Planck's least length, which are named as Treo by Dr. Ashok Saxena in this proposed treo model.

Pure mathematical models, without a true working physical model in hand, drifts in spaculations. The correct approach is that we should have a physical model in hand which is scientifically and mathematically consistant and then its predictions should be tested by the application of mathematics.

Conventional physics deals with HOW and WHAT of any phenomenon but leaves WHY to be answwered later on by some philospher. This work fullfills this basic need of physics and all mathematics which deals with the description of our universe.

I wish him success and hope others would find due corroboration with his model in their own work. I Wish him good health.

Preface

'I want to know HIS thoughts (of creation) rest are details', Albert Einstein.

The basic quest of we humans is to know, WHO I AM & WHERE AM I ? To get the answer of these questions we adopt a spiritual way; the other route is to explore the universe in terms of God's units i.e. Planck's units to DECIPHER HIS SCRIPT written every where.

Prominent thinkers have been convinced that the world observed through our senses represent only the surface manifestation of a deeper hidden reality, where the answer to the great question of existence should be sought. Founding assumption of science is that, the physical universe has coherent **scheme of things.**

The ancients were right, beneath the surface complexity of nature lays a hidden subtext, written in a subtle mathematical code. This 'cosmic code' contains the secret rule on which the universe works; an abstract order which cannot be seen, heard or felt but **deduced.**

With its inborn limitations, we find traditional physics ending up in a blind lane, with no path ahead. From long time and especially in last century all physicists following traditional physics tried a lot, but could not find the answer of most fundamental questions about our universe, few of them are–

A. 1. How to **put all four basic forces (EM, Weak, Atomic & Gravitational force) together in one frame ?**

2. We still want to know, **'what is gravity'?**

3. How to put micro and macro physics together in one frame ? or in other words, **reconcile two pillars of twentieth century – general theory of relativity and quantum mechanics – in a unified field of quantum gravity.**

B. What is **true picture of energy ?**

    1.   What is energy?

    2.   What measures **'one quantum energy'**?

    3.   How **potential energy is converted in to kinetic energy?**

    4.   Whether much sought **fifth dimension of universe, is Energy?**

C. Do we actually understand, Newton's third law of motion, *'every action has equal and opposite reaction'* (Principia Mathematica: 1687), which we all are cramming from last 333 years like parrots. **Can we answer 'what reacts and how'?**

D. So far we have not been able to find **'theory of everything'** as we had no clue about 'the principal which governs our universe'.

The problem with traditional physics is inborn as **we try to study our universe in man made arbitrary units, which complicate our task.**

'Meter[1] used to measure three dimensions of space (length, depth and breadth) is *one metallic rod of fixed length kept in Paris.*

---

1   Base unit of Length, is ten millionth of distance from equator to North Pole of earth, along meridian which passes through Paris, and it was measured in 1793, by French astronomers Delambre and Mechain: the metallic rod of this length is kept in double locked drawer in Paris observatory.

Similarly the weight (mass) is measured by one chunk of metal named as 'kilogram' ("Le Grand K" which is a *piece of platinum iridium*)[2], while we measure time in 'second'.[3]

When these SI units are used to express the value of all universal constants, very less information about our universe is divulged, **and they fail to describe our universe.**

In the study of 'fields of all forces' by pioneers of physics (Newton, Faraday, Maxwell, Einstein, yang and mill & Higgs) arbitary units were used which complicated the matter, and all our physics formulas are still based on them. Because of this in traditional physics, our universe is projected as mysterious place and cumbersome to explain. So any one trying to study universe ends up in a very confused state. **Hence: it is best time to turn around and get out of this blind lane, to search a new path.**

The God has written the script of creation, in 'Cosmic language'. Although the alphabets of this language were discovered, way back in the year 1900 by Mr. Max Planck, but the language has not been drafted till now. These alphabets are Planck units or Gods' units which are universal units, and will never change (some say that even the aliens will be using these units). **But Mr. Planck could not identify, that what these units denote?** The cosmic code is very simple, as HE has written it in cosmic language everywhere very neatly, preciously and meticulously.

Newton, Galileo and other early scientists believed that by exposing the pattern woven into the processes of nature, they truly were

---

2   The standard one kilogram is an arbitrary metallic piece (platinum 90% – iridium 10%), which is kept at the international bureau of weights and measures laboratory at Sevres, France since 1889. By the way, on 19 may 2019 they adopted Planck unit for one Kg, but this measure remains the same.

3   Second is 9 billion (9192631770) oscillations of Cesium atom.

glimpsing the mind of God, and their elegant mathematical form as a manifestation of God's rational plan for the universe.

The fact that the physical world conforms to mathematical laws, led Galileo to make a famous remark 'The great book of nature can be read only by those, who know the language in which it is written and the language is mathematics'. English astronomer James Jeans once said 'the universe appears to have been designed by a pure mathematician.'

The normal work of a theoretical physicist is to investigate an unsolved problem about a natural phenomenon, by applying the laws of physics in the form of mathematical equations and then try to solve the equations to see how well they describe the real thing. By applying the equations that express the laws relevant to the problem of interest, the theoretical physicist can predict the answer.

But there are inherent limitations with pure mathematical models as explained by Gödel's incompleteness theorem [Ref 2]. **Pure mathematical models, even if they deliver correct picture, but without a true working model in hand, all the results obtained will look absurd.** We should have a model in hand and then its predictions can be tested with the application of mathematics.

Here in this, proposed treo model, we will investigate even more basics of the universe, beyond the domain of mathematics, to find out *'where do the laws of nature come from'*.

Space is not empty, but it is permeated by a non-zero energy field, earlier named by Mr. Einstein as 'Space-Time' and by Mr. Peter Higgs the 'Higgs field'. In this model omnipresent interwoven Space – time – energy is named as 'Space matrix'.

Since the time of Newton, when some of properties of this Space (matrix) were first described by his 'three laws of motion' and by his 'gravitational field equations', scientists are trying to understand some more properties of this Space Matrix.

The universe is made up of 'omnipresent, omnipotent ten-dimensional space matrix'. If you want to know its geometry, you will have to analyse the behavior of space matrix, to confirm existing theories and to get answers of still all un – answered questions.

All photons and all mass energy packets up to one unit mass, exert a uniform load on space matrix along its length of spread in a line (on its wave length). This exerted load is neutralized at each apex bound treo by one kinetic coloumn in a Action-reaction mechanism; of Space matrix; to support this packet.

While multiple unit mass cosmic body exert a load by square of unit masses at its gravitational centre which is supported by equal number of gravitons in its gravitational sphere (formed around its one gravitational centre). Where as graviton coloumns of all gravitons at outer most layer of gravitational sphere jointly form its gravitational field.

All bodies are pushed from all possible sides for being supported; once any body comes in gravitational field of another body, both bodies get inadequate support (from common shared space matrix in between), and then both fall towards each other, and we perceive this fall as gravitational attraction of body (Fig. 65, page 221)

The approach of this work is based on **'model dependent realism'**. The proposed Treo model is a 'mathematical consistent model' which is based on direct and indirect observations on our universe and it has been found to be consistent with all known scientific facts. It does not contradict them at most of places, but only explains the basis of these phenomena in a different perspective.

This book is for common man and explains the basic theme and relevant findings of treo model in short. The observations which support this model were already described in detail earlier in my previous books 'Inside a Wave' (Manas Prakashan, 2005) and 'Our Universe and How It Works: Quantum Gravitation and Fifth Dimension'

(Manas Prakashan, 2015; ISBN: 978-9-35235-003-2) and in 'e book' with the same title 'Our Universe and How It Works: Quantum Gravitation and Fifth Dimension' (Book Baby, 2016; ISBN: 978-9-35235-065-0).

For the benefit of all readers, I have described all observations and its inferences in a simple language, while certain concepts are explained with diagrams. Only simple calculations like multiplications and divisions are used in this book, as required to prove this model. The **constant 'C'** used in my first two books is renamed as **constant 'S'** in third book and in this book. There are few unavoidable repetitions for better understanding of the foundation of this new concept. **Already established concepts are written in *italics* while normal font has been used to explain 'Treo model'.**

The understanding of this pear reviewed basic treo model, will not only simplify the teaching of physics, but also help to make testable predictions, which will help in advancement science. **Such a model would be a powerful new tool that will accelerate the rate of new discoveries and inventions.**

'Model dependent realism'

*In 1931 Kurt Gödel, proved his famous '**incompleteness theorem**' about the nature of mathematics. **It showed limitation of mathematics, that there are problems that cannot be solved by any set of rules or procedures** [Ref.2].*

Mathematical models, 'string theory' and 'super string theories' are good but to understand the outcome of these theories, you must have beforehand a clear problem or question on which to apply this mathematics. Even the clear and useful results of M theory, like P2 – two dimensional membranes, or P3– three dimensional blobs of M theory, appears useless without a clear picture of one model. Without a guiding physical picture, a pure mathematical concept sometime drifts into speculations.

Our scientist community proceeds by only describing 'what' and 'how' things work, but 'why' is always left for some later understanding, by some philosopher.

Mathematics will not give insight, which is the work of a philosopher. But a good philosophical model should be a real, fully consistent model (based on physics), which can be written in the language of mathematics.

Until the development of modern physics, it was general opinion that all the knowledge of world could be obtained 'through direct observation', that things are what we perceive through our senses. But modern physics clashes with every day experiences.

To deal with such a paradox we can adopt an approach that is called **'model dependent realism'**. It is based on the idea that our brain interprets the input from our sensory organs by making a model of the world. When such a model is successful at explaining things, we assign it the quality of reality or absolute truth. But there could be many different ways in which one can model the same situation by employing different fundamental elements and concepts and we are free to adopt which ever model is more convenient.

The proposed logistic–mathematical model introduces **one more dimension "Energy" as fifth positive dimension of universe, as part of our** ten dimensional 'Space Matrix' (i.e. space–time–energy of universe; with five positive dimensions formed by **treo** and five curled up negative dimensions represented by **void**).

This book is a small indicator of TRUTH, the truth of nature, and it is a must read book, for any body who want to know answer of iternal questions WHO AM I and WHERE I AM. New observations will further help to describe finer details of working of space matrix and will decipher a new pattern, THE COSMIC CODE based on S number (a new dimension less constant of 1.85539×10^{43} number) which is Rhythm, at which universe is vibrating at planck frequency.

Acknowledgements

I will like to thank all my reviewers, espacially to Dr. R.C. Saxena senior scientist from NPL Delhi, and Dr. Avinash mathur retired as principal scientist from CSIR, as both guided me and kept a constant watch and frequently reviewed my work time to time.

How can I forget the contribution of Mr. Ajay Singh professional artist at Agra, with whom I made all diagrams and cover pages of last 3 books. My both sons Er. Ashwarya Saxena and Er. Amit Saxena remained my great critic and with their concrete views and timely support, I could deliver this work.

Dr. (Prof) Surekha Saxena, my wife deserves special thank for *having faith in my work* and for untiring efforts made to shift this model from my brain to paper, understandable by common man.

Most greatfully, I am thankful to all mighty God, who have provided me time, energy, passion and power and even trained me to use my subconscious (90% mind always remain unused) to make all important observations and to perform all 'thought experiments' in depth, to accomplish *his assigned task*.

CHAPTER 1

Story of Creation

'Nothing comes out of nothing'

This model advocates that **only 'one object' and 'one tool' is used in Creation**[4]. But what is this **one object** and how this **one tool** is used to carve entire creation, are subject of this book.

This book details a new quantum model, named the 'Treo model', which uses Planck's units and explains beautifully the structure of universe and its working.

1. Treo as Precursor (Generator) of All Space and All Matter

The proposed model presents a hypothesis about a fundamental primordial particle commonly known as a *string,* **which is 'Treo' in this model**. Treos are one-dimensional and each is of **Planck's least length**.

Treo present themselves only in one dimension of length, have a fixed mass and energy and they vibrate continuously, each time in different plane, to create all five-positive dimension of space represented by active space–matrix.

---

4   Occam's razor by William of Ockham (1287-1347) [Ref 1] is a problem-solving principal – it dictates that in a model, fewer the assumptions made, the better is the model.

Size of one treo = *Planck least length* = 1.616229×10^{-35} meter

Weight of one treo = $6.32582652229 \times 10^{-95}$ kg (See page 218)

Energy of one treo = $5.6831651 \times 10^{-76}$ J (Applying $E = mc^2$)

We will first see that how space–time–energy and matter is created, by this Treo.

Treos are divided in two categories 'free treos' and 'bound treos'.

Quanta of **free treos**; accumulate and pile up in integral numbers to produce mass energy of all matter and photons; while **bound treos are arranged alternately with voids,** to construct the geometry of all three dimensions of Space, marks fourth dimension of Time, while they themselves are vibrating particles by which they represent fifth positive dimension of universe; ENERGY.

2. Bound Treos

Five positive dimensions are created when they arrange themselves and oscillate at **Planck frequency (S times per second),** each time after a gap of Planck's least time, in all possible S planes. Bound treos after deformation account for all DARK MATTER of universe.

The rhythm of vibrations of all bound treos in space matrix, at Planck frequency regulates our universe.

Planck frequency = (1/Planck Least Time) = $1/0.5389689 \times 10^{-43}$ sec = $1.85539441 \times 10^{43}$ per second

This number $1.855394409 \times 10^{43}$ is S; the proposed new dimensionless constant in this model.

All bound treos vibrate by S number of vibrations per second in all possible S number of planes.

The constant S determines several other universal constants.

1. **S number of treos is in one quantum energy**, which is also the 'quantum of unit action' and the value of *'reduced Planck constant'; ħ (h bar).*

2. While *Planck constant h:* is the **angular momentum of this one quantum mass–energy** i.e. **h = S × 2π.**

3. **S bound treos distance per second** (in *S* vibrations) **is the** *Speed of light, c.*

4. **S free treos is one quantum mass energy;** and are present in *one-unit photon.*

5. **√S quanta are mass energy of** *one-unit electron*; and it is *one-unit charge.*

6. **S quanta or S² free treos makes one-unit mass (*Planck mass*).**

7. **One-unit mass is supported by S² kinetons per second per second**, it is also the **derived value of *Gravitational constant*.**

8. While uncurling of each void per vibration, by **1/S² of one bound treo length, at each point of universe,** is the value of *cosmological constant* and the rate of expansion of universe, and it calculates the present value of *Hubble constant.*

9. S seconds (S² vibrations) *is one life span of universe* and S² bound treos forms *the radius of universe (presently contracted in 13.8 biillion light years.)*

3. Creation of Space

Space is not empty, but it is permeated by a non-zero energy field, earlier named by Mr. Einstein as 'Space-Time' and by Mr. Peter Higgs the 'Higgs field'. In this model omnipresent interwoven Space – time – energy is named as 'Space matrix'.

All Bound treos are placed alternately with voids (voids have five curled up negative dimensions) and thus it weaves our ten–dimensional uniform, omnipresent space–matrix.

These one-dimensional treos can form 3 dimensional space, by their mode of compilation, when arranged side by side along a line, it produces first dimension of **Length**.

To construct two dimensions of **length and breadth**, they construct figure of squares.

Finally they construct all 3 dimensions of space (**length, breadth and depth**) small cubes of Planck least length are formed in multi layered space matrix.

One unit space matrix is a big cube where length of its **each side is of S bound treos**, Fig 12 page 85.

4. TIME, The Fourth Dimension of Universe

Time along with space both together form four dimensions of space–time. **Each bound treo vibrate by** S number of vibrations per second; and time between two such vibrations is **Planck least time**. All the treos of the universe vibrate simultaneously and thus whole universe vibrates at S frequency.

The minimum time required for any action to take place is Planck least time i.e. the time period in between two vibrations of universe. It may be a physical action or any chemical or biological processes which occurs around us continuously, is perceived by us as 'flow of time'. No time machine (to visit past or future) is possible as we move in constantly changing universe, transforming with its each new vibration, and leaving behind no traces e.g. as our body has evolved from a baby, with no remanents of baby body left, so we can not transform our present body in a baby body.

If **there is a reduction in total number of vibrations per second** in any local area, the time between two vibrations (i.e. Planck's least time) will increase, which in turn will increase 'minimum reaction time' or the 'least processing time' (now every action or chemical and biological process will take more time), and thus it will slow down the rate of change, which we perceive as slow down of time in involved local area. [e.g. when food is kept at lower temperature (with expansion of local space matrix) in fridge the local time slows down, which increases processing time of all biological processes of bacteria already present in food, and thus the multiplication of bacteria slows down; and it takes more time to get food spoiled].

If no change occurs Time will stop, then the universe will look like a photograph.

Fourth dimension of Time

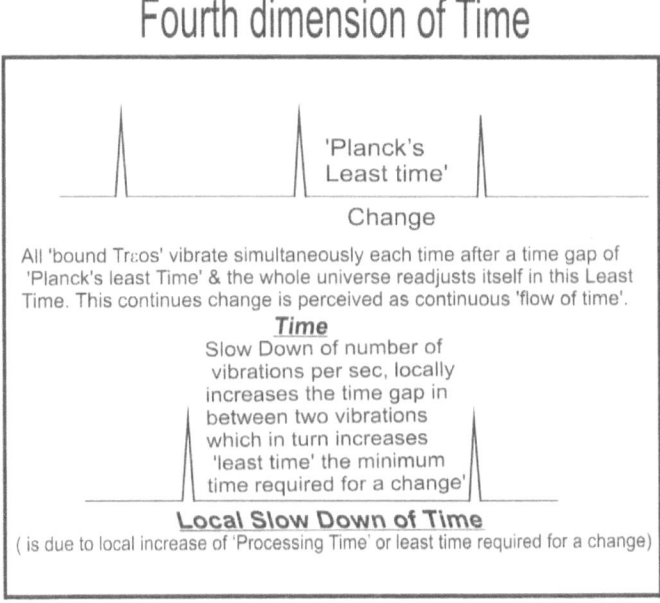

Figure 1: Each Bound treo vibrate by S Times per second, with a time gap of Planck's least time.

Number of vibrations in one second

One second/Planck's least time = $1/ 0.539 \times 10^{-43}$

=$1.855394405 \times 10^{43}$ vibrations per second (S vibrations)

5. ENERGY – Potential and Kinetic Energy

Although Stephen Hawking's remained reluctant with extra dimensions, but Energy as Fifth extra dimension of universe was hailed in a different way, by his remark *'However, there has recently been the suggestion that one or more of the extra dimensions might be comparatively **large or even infinite**. This idea has great advantage...'* [ref 50]

All Electromagnetic energy and all mass energy packets up to one unit mass, exert a load along its length of spread in a line at each apex bound treo on its wave length, Fig 13 page 87.

At each apex bound treo the load exerted by any number of free treos is neutralized by equal number of bound treos of space matrix, as they convert in kinetons and get arranged in kinetic coloumns.

All bound treos of space–matrix, which are not subjected to load, simultaneously vibrate by S vibrations per unit time (per one second), in S number of all possible planes, Fig 2.

When any one such bound treo vibrating in all S planes is subjected to (minimum) 'load' of one free treo, its *planes* of vibration (or the degree of freedom) is restricted and this bound treo will now vibrate S times per second only in one plane, in the direction of load to support or neutralize this load (mass pressure or momentum) on space–matrix, by an equal and opposite reaction, Fig 3. **Thus contracted bound treo, vibrating only in the direction of load, is a 'kineton'.**

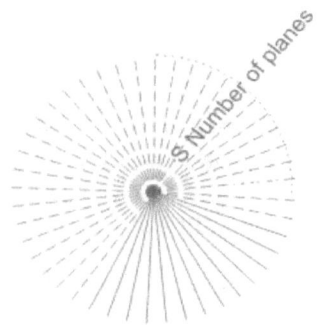

Figure 2: Bound treo vibrates in all S planes or has S degrees of freedom, completing S vibrations per second

Load of one free treo

One bound treo starts vibrating only in direction of load

Figure 3: When bound treo of space–matrix starts vibrating only in the direction of load (minimum of one free treo) to support it, this supporting bound treo deforms and is called a 'kineton'.

6. Action Reaction Mechanism

We throw a ball on wall and it comes back to us, a rocket accelerating in space releases hot gases back from its engine, an action, and in a reaction it gets a forward thrust which pushes rocket forward, or while walking when we push earth back ward by our foot, we get a forward push to walk.

These are few examples of Action-reaction mechanism governed by third law of motion 'every action has equal and opposite reaction', but we are still unaware about, **how** this action – reaction mechanism is executed in universe.

Figure 4: Action – reaction mechanism propel rocket in space.

For any increased 'load' (any mass or momentum) of any number of free treos, **equal** number of kinetons arrange themselves according to coloumn geometry (as discussed below, on page 86) to neutralize this load, and thus Action-reaction mechanism is executed by space matrix, Fig 4.

The space-matrix has three interdependent components of **space-time-energy,** where if one component changes its value, the value of other two components will accordingly change instantly. When space gets locally contracted by conversion of bound treos in to kinetons, the potential energy of each involved bound treo will be converted in to kinetic energy.

The vibrating bound treos in all possible S planes in undeformed space matrix generate potential energy of universe (*so undeformed space matrix has nonzero energy in its field*). But when the load on space matrix restricts the plain of vibration of equal number of involved bound treos, which now vibrate only in the direction of load (to support it indefinitely) to be converted in to kinetons; thus they contract local space and generate kinetic energy. In execution of this action-reaction mechanism in this quantum universe, with increasing load all three dimensions of **one-unit space matrix** gets involved, one by one.

All free treo packets of all photons and all elementary particles spread evenly on space matrix in a line and exert uniform load (Action) on each apex bound treo along its length (wave length), according to its packet density.

This exerted load at any one apex bound treo (or at one gravitational center of multiple unit masses body) is supported by one kinetic coloumn.

With ever increasing load, the deformation/contraction increases, and equal numbers of bound treos are converted in kinetons. After upper limit of first dimension is crossed, second and then third dimension of same unit space matrix gets deformed.

The contraction of unit space matrix, first involves all (**S**) bound treos in length, but still increasing load gradually involves all (**S²**) bound treos present in one sheeth (length and breadth) of unit space matrix, and finally with contraction extending in depth of unit space matrix as well, will involve all (**S³**) bound treos of one unit space matrix.

Thus, by complete deformation of first (length), second (length and breadth) and all three dimensions of space (of one-unit space matrix), the required energy of each dimension is achieved (which is the energy of S kinetons, S² kinetons and S³ kinetons respectively).

Kinetons represent kinetic energy and S number of kinetons is **one quantum kinetic energy.**

7. Speed of Light

(Speed of displacement of photon packet per second)

Each time after one Planck's least time, by **next one vibration, the photon packet is pushed to adjoining 'next bound treo'** in direction of its progression on space matrix. And thus by S vibrations in one second, it moves **by S number of (i.e.$1.855394405 \times 10^{43}$) bound treos distance per second.**

$1.855394405 \times 10^{43}$ vibrations per second × $1.615788303 \times 10^{-35}$ Meter[5] is displacement of photon packet per vibration = **2.997924×10^8 meter per second, and it is conventional value of 'c', the Speed of light.**

This calculation indicates, (as it is calculated from two constants of Planck's least length and Planck's least time), that the speed of light is also a constant and is ultimate speed of movement in our universe, in un–deformed space matrix. The speed of light also confirms the proposed 'structure of space matrix' as described in this treo model.

---

5    $1.615788303 \times 10^{-35}$ Meter is Space occupied by one bound treo in length.

Thus, space matrix is a medium, which propels all photons, and every motion of all bodies in this universe is the result of 'propulsion by this active medium': It is to be noted, *the continuous motion of every thing in this universe,* **is due to this active space matrix.**

As this moving deformation can not die by itself (with out being opposed by other deformation) and thus it is conserved and thus photon moves indefinitely; it is the basis of conservation of momentum and the reason behind first law of motion (*if a body is moving it will go on moving indefinitely until a retardation force is applied on it*)

8. Matter – All Matter Is Formed By Integral Number of 'Cmue'

(35.012 MeV; Which Is 'Composite Mass Of One Unit Electron')

All types of Photons are formed by Integral multiple of **one quantum mass energy (S free treos).** There are √S quantum levels in each dimension of space –time. The free treos, are added in unit of one quantum i.e. S free treos (present in **one unit photon**), at each of √S quantum levels of first dimension of length to produce all √S types of photons which forms complete EM spectrum.

Then the increments occur in unit of **one-unit electron** made up **of √S quanta mass energy (i.e. S × √S free treos)** at each of √S quantum levels of second dimension (of length and depth) and thus it produces all elementary particles, nucleons, atoms, molecules and up to one–unit mass (*i.e. Planck mass= 2.176 × 10⁻⁸ Kg*), at their matching quantum level. All big bodies are made up of multiple number of unit masses.

But when it comes to pack **charged unit electron packets togather** (which may be visualized as one 'brick', Fig 5), to form a packet of composite elementary particle or nucleon, **great amount of localization energy** is required, (which may be visualized as 'cementing material',

Fig 6) to overcome the strong repulsive forces in between two negatively charged unit electrons.

Figure 5: One-unit electron (as one brick) is building block of all matter

Figure 6: Unit electrons (bricks), In different numbers along with its localization energy (cementing material), form all elementary particles.

Author is first person to observe and document it for the first time in his previous book [Ref 6]; that 35.012 MeV is **Composite mass of unit electron; is** *the mass energy of unit electron* **alongwith its localiztion energy.**

All matter (mass energy packets of all elementary particles and neucleons) is produced with integral number of composite mass of

unit electron (Cmue; prononce − − C mue) and are placed at their respective matching quantum levels in second dimensional deformation.

Bound treos of space matrix are the source of this localization energy.

While describing the Higgs fields it was advocated by Mr Peter Higgs, that Elementary particle drag the Higgs field (i.e. space matrix made up of bound treos) with them in which they move; by **Higgs mechanism.** Examples are.

    a.   *Protons are composite particles, most of Proton mass is not only its mass energy but predominantly its quantum chromodynamics binding energy which is attributed to gluons.* These are bound treos (kinetons) contributed by space matrix (as localization energy) or the wrapped Higgs fields. **27 Electrons along with their localization energy form one Proton.**

    b.   *The known fact that 'mass of any fast–moving particle increases with its increasing velocity' (one of the postulates in theory of relativity),* is due to transfer of kineton layers (along with its riding free treos) in all kinetic coloumns of moving body, from all kinetic coloumns of pushing body.

The formula for calculating, the composite mass of unit electron which is named as Cmue is discovered by me and was earlier published for the first time as 'mass units' in my previous book [Ref 6]

The fine structure constant is coupling constant and is ratio of energy required to counteract the force of repulsion in between two electrons vs. reduced compton wave length of photon.

According to proposed formula, the Composite mass of unit electron can be calculated by dividing the mass energy of unit electron 0.511 MeV by twice of 'fine structure constant', with its known value of 1/137.035999084 [Codata 2018 value, Ref. 51].

> Composite mass–energy of unit electron (35.012 MeV) = Mass energy of one unit electron/2 × Fine structure constant
>
> 35.0126237 MeV = 0.510998918 MeV/(2 × 1/137. 035999084)

All elementary particles and nucleons have their mass energy which is integral multiple of this Cmue composite mass energy of unit electron (mass energy of unit electron with its localization energy), within 1% deviation. (Ref below, table 1)

PARTICLE	Known mass of elementary particles in Mev	Calculated Mass of all elementary particles in Mev	Known mass-calculated mass × 100/ known mass = Error in %	Number of composite unit electron masses in particle as observed here
LEPTONS				
Electron	0.51 Mev	0.51/2 (1/137.035999710) = 35.012698 Mev (Composite mass of unit electron or m$_e$)	0%	= 1
Muon	105.66 Mev	m$_e$ × 3 = 105.3809 Mev	0.26%	= 3
Tau	1776.99 Mev	m$_e$ × 51 = 1785.6475 Mev	0.48%	= 51
MESONS				
Pion	139.57 Mev	m$_e$ × 4 = 140.0507 Mev	0.34%	=4
Kaon	493.68 Mev	m$_e$ × 14 = 490.1777Mev	0.70%	=14
Eta	547.75 Mev	m$_e$ × 16 = 560.2031 Mev	− 0.22%	=16
Rho	775.8 Mev	m$_e$ × 22 = 770.2792 Mev	0.71%	=22
Omega	782.59 Mev	m$_e$ × 22 = 770.2792 Mev	1.57%	=22
Meson	1869.4 Mev	m$_e$ × 53 = 1855.6728 Mev	0.73%	= 53
Ds Meson	1968.3 Mev	m$_e$ × 56 = 1960.7109 Mev	0.37%	=56
B Meson	5279.4 Mev	m$_e$ × 150 = 5251.9023 Mev	0.52%	= 150
Bs Meson	5369.6 Mev	m$_e$ × 156 = 5391.955 Mev	0.41%	= 156
BARYONS				
Nucleons	938.27 Mev	m$_e$ × 27 = 0945.3427 Mev	− 0.75%	N=27
Lamda	1115.68 Mev	m$_e$ × 32 = 1120.4062 Mev	0.42%	N=32
Sigma	1197.45 Mev	m$_e$ × 34 = 1190.4318 Mev	0.59%	N=34
Xi	1314.18 Mev	me × 38 = 1330.4825 Mev	− 0.12%	N=38
Omega	1672.45 Mev	m$_e$ × 65 = 1680.6095 Mev	− 0.48%	N=48
Lamda$_c$	2284.9 Mev	m$_e$ × 48 = 2275.8254 Mev	0.397%	N=65
Sigma$_c$	2452.2 Mev	m$_e$ × 70 = 2450.8889 Mev	0.04%	N= 70
Xi$_{cc}$	2466.3 Mev	m$_e$ × 70 = 2450.8889 Mev	0.62%	N=70
Omega$_c$	2697.5 Mev	m$_e$ × 77 = 2695.9778 Mev	0.05%	N=77
Lamda^e$_b$	5654.0 Mev	m$_e$ × 161 = 5637.0445 Mev	0.29%	N=161

Table 1: Mass–energy of elementary particles is integral multiple of composite mass of unit electron within 1% deviation

While forming a nucleus by protons, neutrons, mesons, gluons and W and Z particles, **this localization energy** (wrapped space matrix as cementing material with each unit electron) **gets partially converted in to binding energy of nucleons,** which is manifested as 8 types of Gluons (type of bosons made up of kinetic coloumns).

The localization energy, binding energy (gluons), and supporting kinetic coloumns (of W Z boson) all three are the components and source of Atomic energy.

The number of 'free Treos' calculated in [Ref 6].

1. Unit Photon = $1.855394405 \times 10^{43}$ Free Treos

2. Gamma photon of 1.02 Mev = 2.8724704×10^{64} Free Treos

3. **Electron of 0.511 MeV** = $1.439491604 \times 10^{64}$ free treos

4. *Proton* = $2.64416818 \times 10^{67}$ free Treos

5. *Neutron* = $2.64781297 \times 10^{67}$ free Treos

6. Unit mass (*Planck mass*) = $3.442488398 \times 10^{86}$ Free Treos = or *2.17643×10^{-8} Kg (Planck mass)*

7. **One Kg mass** = 1.5808523×10^{94} Free Treos

8. Number of treos in one unit charge i.e. one eV = $1.439491604 \times 10^{64}$ free treos

    (charge on one treo = 1.112167×10^{-83} coulomb; and one coulomb charge is on $0.899240188 \times 10^{83}$ free treos, *in 6.2 × 10^{18} unit electrons*)

9. Number of treos in one calorie (*calorie = 4.184 joule*) = $0.736362761.6 \times 10^{78}$ free treos.

10. *Charge mass ratio = 1.76×10^{11} coulomb/Kg* [$1.76 \times 10^{11} \times 0.899240188 \times 10^{83} = 1.58(26624) 10^{94}$ free treos]

    Similarly, by using these values, number of free treos can be calculated in any type of photon, in all type of elementary particle and in any element.

9. Expansion of Space-Matrix (Basis of 'Hubble Constant' and 'Cosmological Constant')

(It is again reminded; that all text written in italic in this book are already known concepts while text written in normal font is advocated by proposed Treo model)

Some spectacular advances in observational cosmology of twenty first century are, detailed mapping of the cosmic afterglow of the big bang which was done by satellite named WMAP and the new discovery of **accelerating expansion** *of universe by some mysterious dark energy as analyzed by the data collected by Hubble space telescope. But what is this dark energy and how it works?*

(a) Our space is not empty.

Stephen hawking argued *"The value of field and its rate of change play the same role as 'position' and 'velocity of a particle'. The 'Heisenberg uncertainty principal' dictates that more accurately one is determined less accurately other can be. Empty space means that both the value of field and its rate of change are exactly zero. Since the uncertainty principal does not allow the value of both the 'field' and 'rate of change' to be exact (i.e. zero), space cannot be empty".*

It can have a state of minimum energy (a non-zero energy state) called 'Vacuum', but this state will be subject to vacuum fluctuation. This is also termed as Quantum foam out of which virtual particle pairs appear and then annihilate each other. The virtual particles have energy so there should be infinite amount of energy which should curve (according to general relativity mass/energy both curve space–time) and contract the universe to infinite small size. But this does not happen, so what prevents it?

'Cosmological constant' was brought forward by Einstein to explain the model of a 'static universe', but when it was proven that our universe is an 'expanding universe', he admitted that cosmological constant was the

*greatest blunder of his life. But now days, the **cosmological constant has again emerged, and being projected as some sort of 'dark energy', which prevent universe from collapsing.***

(b) *The universe is expanding like a balloon.*

Let us see, **the role of negative energy incorporated with all five negative dimensional curled up 'voids', all of which are continuously uncurling?**

There could be only one possible explanation for the 'balloon like expansion of the universe' that the **universe is expanding at its each and every point,** with 'constant flow of time' and both the expansion of the universe and flow of time are based on a single mechanism.

Simultaneous uncurling of all the voids leads to balloon like expansion of universe with passage of time, thus the present size of universe also indicates present age of universe.

The space occupied by each 'void' (**its size**) is always increasing due to simultaneous uncurling of its curled up five negative dimensions, @ **"$1/S^2$ of present size per vibration" and this is also the value of "cosmological constant" and mechanism of accelerating expansion of universe (according to treo model).** This expansion is by uncurling of <u>one</u> spherical spike, originating from any one of S dots in S planes and S dots from S directions, as total curled up S^2 pikes or imaginary empty spaces forms one void.

It is to be noted here, that the **derived value of gravitational constant is 'S^2 kinetons per unit mass per second per second';** (see calculations in the section of gravitational constant page 218).

If the radius of our universe is 'S^2 pairs of bound treo and void' as postulated/calculated in this Treo model (also see below the Hubbell's length), then by this combined and simultaneous uncurling of all these voids at each point on its radius, the radius of the universe at its periphery will increase by **one Planck's least length per vibration.**[6]

Thus our treo model explains why *the periphery of universe will always swell up, with the speed of light (i.e. by one Planck least length per vibration). Therefore size of present universe is measured in light years; and its present size is 13.799 billion light years.* This means the balloon (at periphery) of universe is inflating continuously at the speed of light.

The **accelerating** rate of expansion of universe can be explained, with increased value of Planck least length with each vibration of universe.

When **each void constantly uncurls itself at each point**, it not only expands the universe but thus, simultaneously and continuously tries to flatten 'whole crumpled deformed space matrix' of universe. The fully 'flat sheath of un–deformed space matrix of universe' is the requirement for big crunch: i.e. death of universe.

This constant expansion leads to decrease in its 'total kinetic energy' and in turn with continuous increase in its 'total potential energy', also results in slow aging of universe (and all its creatures; which are infect 3D printed free treos on space matrix in different shapes).

(c) **To make space matrix flat like a plain sheath with no wrinkles** (deformations/contractions)

But along with expansion, one more mechanism is working to **make space matrix flat and free of local deformations.**

To accomplish this task, all cosmic bodies because of their gravitational attraction first gather all matter present around it.

Then the gravitational spheres of all cosmic bodies e.g. Sun, (which are **black holes in formation** or quasi black holes) *along with unit*

---

6   *Accelerons* are the hypothetical subatomic particles that integrally link the new found mass of the neutrino to the dark energy conjectured to be accelerating the expansion of the universe.

black holes (not singularities but eternally collapsing objects) continuously churn the matter with the production of 'Hawkins energy pair' of positive and negative energy particles i.e. which are recognized as free treos and voids in this model (and are the source of energy provided for outwards flow of 'Solar wind' from Sun and similarly from other cosmic bodies).

Now **all deformations** (kinetic coloumns) which were earlier supporting this matter, before it churned out, will eventually **vanish, to result in more flattened local space matrix**.

By these two mechanisms, total potential energy of universe is increasing, at the cost of decreasing total kinetic energy, with this continuous expansion and flattening of universe; both of which are increasing the 'entropy of universe'.

(d) The end of universe

All matter (mass energy packets) and space matrix geometry, in universe is preserved as long as five negative dimensions of voids are curled up. The continuous uncurling of curled up 'five negative dimensions of voids', at some moment in future, will match with five positive dimensions of treos and then all treos and their full–size adjacent voids will engulf each other instantly (after **one life cycle of universe of S seconds**, on death day of universe). It will result in a big crunch or 'death of universe', and **all space matrix of universe will instantly collapse.**

Then the 'resultant **sudden contraction of space matrix of whole universe'** around one universal singularity; will convert dying universe from 'maximum potential energy state' to 'maximum kinetic energy state' of new born universe. And thus the 'big crunch' is immediately followed by 'big bang', and then newly formed fully contracted universe starts its new life cycle/and next oscillation of five dimensional pendulum of our 'pendulum quantum universe'.

(e) Hubble's law

Our universe is constantly expanding at each point like an inflating balloon. Thus, the distance between any two points in the universe is constantly increasing and all the galaxies are getting apart. It means that the expansion distance in two–kilometer length is twice that of expansion distance in one kilometer length and it will be tripled in three–kilometer length.

This proposed simultaneous uncurling of all voids can explain this type of expansion of universe. The uncurling of each voids[7] at the proposed rate

"$1/S^2$ of present size per vibration" will increase the present radius (S^2 bound treos) of universe by 'One Planck's least length per vibration' which is the speed of light. Thus "one Planck least length/S^2 per vibration" is Hubble's constant H0.

(f) 'Present value of Hubble's constant

(Conventionally Hubble's constant or H0 is used in known formula ($v = H0 \times D$)

Hubble's constant i.e. H0; is 'constant of proportionality', where 'v' is velocity of receding galaxy and 'D' is distance of galaxy.

Rate of expansion of our universe when calculated according to proposed rate, per second per Mega Parse, in present contracted size of universe (of approximate 13.8 billion light years), matches with the **present accepted value of Hubble's Constant, *which is about 70 (+ – 2.4) Km per sec per Mega Parsec*[8].**

---

7   *'Inflaton' described in literature is a hypothetical particle responsible for cosmic inflation.*

8   *one parsec is 3.26 light years distance, or 30.857 pentameters (10^{15} meters); One mega parsec = one million parsecs*

Hubble's Time $1/H0$ = S^2 vibrations ('total life span of universe is S seconds'; from its birth up to the time of its death, at 'big crunch').

While the present age of universe is 13.8 billion light years.

Hubble's length $1/H0 \times c$ = (S^2 vibrations as proposed 'life span of universe' × one Planck's distance per vibration) = S^2 **Planck's distances; will be the radius of our universe at the time of 'big crunch'.**

CHAPTER 2

Forces

---

1. What is Force?

"Force is an influence, which causes an object to undergo certain change, either concerning its movement, direction, or geometrical construction".

All forces which exist in universe are solo or combined manifestations of four basic forces (electromagnetic, gravitational, Atomic & weak forces).

Unification of all these four basic forces is to put all forces together in one frame or to prove that all forces infact are manifestation of only one force of nature and are produced by one single mechanism. **It was earlier tried but was partially successful (as gravitational force could not be unified with rest of three; because their was no exsisting concept about true nature of gravitational force).**

What forms fields of all forces?

For the past 333 years, when equations of gravitational fields were written by Sir Newton, scientists are **searching for the '*field equations.*' of all forces of nature.**

We will adopt a different and even more basic approach for unification of all the forces. It is well known fact that the theoretical physicists sometimes twist the theorem to get the desired result, but a **small twist in our beliefs is required** for unification of forces and to understand the mysteries of our universe.

It was earlier quoted that 'Force is nothing but geometry of space'.

According to proposed treo model, fields of all the four forces are produced by reacting space–matrix in response to the applied load. The four basic forces generate as a result of **'load dependent Action –Reaction mechanism of Space matrix in increasing number of dimensions of space and time'**. Each force is limited to deformation in one dimension and changes its nature, with succesive involvement of increasing number of all four dimensions of space–time, and thus acquires its unique and changed geometry of all four forces.

2. Relevant Back Ground History

*The universal mathematical language of theoretical physicists is '**field theory**'. Fields were first introduced by Michael Faraday, son of a poor blacksmith. Faraday was a self–taught genius who conducted elaborate experiments on electricity and magnetism. He visualized 'Lines of Force' emanated from a magnet and electric charge in all directions, which filled up all space. With his crude instruments he could measure all around, the strength of these lines of force, arising from a magnet or an electric charge. Thus he 'assigned a series of numbers to these points' about direction and strength of these lines. **Field, is a collection of numbers defined at every point in space, that completely describes the force at that point.***

*Initially on June 10, 1854 AD, Bernhard Riemann in his elegant essay 'On the hypothesis which lie at the foundation of geometry', introduced the world for the first time, to the dazzling **properties of higher–dimensional space** and toppled the pillars of classical Greek geometry (Euclidian geometry BC 300).*

***Euclidian geometry is the geometry of two–dimensional plain surfaces where Pythagoras theorem is obeyed.** It states that the square distance between two points in space is the sum of the squares of its perpendicular components.*

And

> *– Sum of three angles of any triangle is equal to 180$_0$*
>
> *– Two parallel lines when extended never meet*
>
> *– Shortest distance between two points is a 'straight line'.*

Non–Euclidian geometry as proposed by Mr. Riemann, is the geometry of higher dimensional *curved surfaces*

where

> *– Sum of three angles of any triangle is more/or less than 180o*
>
> *– Two parallel lines when extended will meet at some point*
>
> *– Shortest distance between two points on a curved surface is not a straight line, but the 'arc of its great circle.'*

Riemann concluded that electricity, magnetism and gravity is caused by ***crumpling of our three–dimensional world in fourth dimension.*** *Thus, the force has no independent life of its own, it is only the apparent effect caused by the distortion of geometry; or the* ***'force was a consequence of geometry'.***

(He had several nervous break downs, as he was suffering from bipolar disorder and died poor with tuberculosis at the age of 39).

The Faraday's fields were plain two–dimensional surfaces described by two numbers. For curvature of this surface, we need one more (third) number to describe a curved field.

But in four dimensions or in any higher 'N' dimensional surfaces, we require 10 numbers or 'metric tensor's' at each point to describe it, as used in 'general theory of relativity', by Mr. Einstein.

It was advocated that, the Secret of unification lies in expanding, Riemann's metric to N dimensional space and then to cut it up in rectangular pieces. Each rectangular piece corresponds to different force.

Further work was left for Maxwell and Einstein. 1n 1860 **James Clark Maxwell wrote down equations for electric and magnetic fields.** **Mr. Clark Maxwell** *with his equations succeeded in the unification of two quite different forces of nature i.e. electricity and magnetism as one & he could also explain the light as Electromagnetic wave.*

One can think of Maxwell field as being made up of waves of different wave length. In a wave the field will swing from one value to another like a pendulum. But it should be noted that the ground state or lowest energy state of a pendulum must not have zero energy.

The new 'non–Euclidian' Riemann's geometry was later used by Einstein in his 'general theory of relativity' to describe 'gravity' in four dimensions of Space–Time. Though his 'metric tensors' gave him a powerful way to describe a curved surface in any dimension, but initially Einstein did not know, the precise equations that the metric tensors observe i.e. what made the sheet crumpled. One of Einstein friend gave him precise equations, by which he could carve the 'general theory relativity' to describe gravitational field.

While **Mr. Schrödinger** *with his 'Schrödinger's equations of quantum mechanics describes the whole 'standard model of elementary particles' and 'structure of atom'.*

In 1970 the field equations (yang – mills field) for 'subatomic forces' (weak forces) could finally be written down by utilizing the earlier work of C. N. Yang and R. L. Mills.

How light travels trillion of kilometers through the vacuum of outer space? Experiment showed beyond question that light is a wave, and then it will require something 'waving'. All waves like sound waves or water waves require air and water respectively the medium which waves. Then how can light be a wave, if there is nothing to wave?

Michelson's & Moor's experiment only proved that there is 'no relative motion' of Earth in ether. Thus, the concept of 'relative motion of bodies in ideal ether', & 'ether' itself vanished, but the concept of medium was never denied.

Einstein in his 'general theory of relativity', advocates space–time a medium which get deformed in presence of a cosmic 'body' e.g. Sun.

The Kaluga–Klein theory

This theory was proposed in 1923, and it advocates presence of a fifth dimension. **(But this theory could not identify what is fifth dimension of Universe. In this proposed model for the first time ENERGY has been identified as Fifth positive dimension of universe.)**

*This theory gives the simple explanation; if light could travel through a vacuum it was because the **vacuum itself was vibrating** in four dimensions of space–time[9].*

The advantage of any 'hyperspace theory' is that the 'Maxwell fields', 'Einstein's fields' and 'yang mill fields' can all be placed easily in one 'hyperspace field theory'.

The laws of nature become simpler and elegant in higher dimensions. In higher dimensions we have enough room to unify all known forces. But we cannot visualize higher dimensional spaces. Our brains can only perceive up to three dimensional spaces. (How can you describe the color of a rose, to a person who is blind since birth?)

By adding fifth dimension the Kaluga and Klein proposed a new theory of gravity in which the light could be explained as vibrations in fifth dimension.

---

9   **Mass dependant deformation or some type of Coloumn like structures were even earlier described in K K Theory.**

The KK theory describes '**Kaluza Klin towers**': For example, on the simplest of principles, one might expect to have standing waves in the extra compactified dimension (**in proposed model this is fifth dimension of energy**). If a spatial extra dimension is of radius R, the invariant mass of such standing waves would be $M_n = nh/Rc$ with n an integer, h being Planck's constant and c the speed of light. This set of possible mass values is often called the **Kaluga–Klein tower**

Thus, the force of gravity and light can be united in a simple way by describing the gravity in four–dimensional space–time and keeping one dimension reserved for electro-magnetic forces out of five dimensions.

But the theory had many difficult technical problems which rendered it useless for half century. More advanced versions are 'super gravity theory' and later on 'super string theory' which eliminated the inconsistencies of the Kaluga & Klein theory.

'Super string theory' and 'all M theories' (and this model) *postulate that all matter consists of tiny vibrating strings. The length of one string is 'one Planck's least length';* **which is named as treo in this proposed Treo model.** *Presently M theories and all 'Super string theories' use this higher dimensional geometry in ten dimensions, in an attempt to lay down laws of physical universe.*

The 'Super string theories' calculate by assuming that our universe is in ten dimensions. *It is broken down in two parts 'one four dimensional our present universe', while the rest of six dimensions 'are curled up' in 'Planck's least length'.*

The M theory is not a theory by itself, but it is a collection of many theories in which each is a description of observations in some range of physical situation. The different theories, in the M family may look different but they all show the same underlying phenomenon.

All 'M theories' and 'Super string theories' and the 'proposed Treo model in this book' all advocates this higher dimensional geometry in ten dimensions, in an attempt to lay down laws of physical universe.

Elephant of 'Quantum gravitation'

"Five blind persons those who have never seen or heard about Elephant, were asked to describe this animal by its feel. The first person holding its leg described Elephant **'like a pillar'**; the second person examined its stomach and found it **'like a bag'**. The third person examined its tail and declared elephant is a **'bottle brush'**, while the fourth at its ear described it to be a **'sheet'** and the fifth blind person after examining the trunk of elephant declared it a **'pipe'**. You can notice that none of these descriptions describe the elephant."

Same is the story 'about our concept of 'Gravitation'. Most of scientists while attempting to define gravitation considered only one aspect of phenomenon as in Elephant story (i.e. Gravitational attraction by Newton or Slope of space time by Einstein) and therefore over looked the full picture of Gravitation.

In his story of gravitation, Mr. Newton (Sir Isaac Newton 1643 to 1727) described Gravitation as 'a force of attraction between bodies' which depends on their masses and distance at which they are placed. He calculated (1686 AD in his book principia) the value of 'Universal Gravitational constant'. Presently it is described as a 'force of attraction between two masses of one Kg each, which are kept one meter apart' and the 'force of attraction is **inversely proportional to the square of distance'.**

These were first correct observations by which we people on Earth could get the first glimpse of the working of heavens.

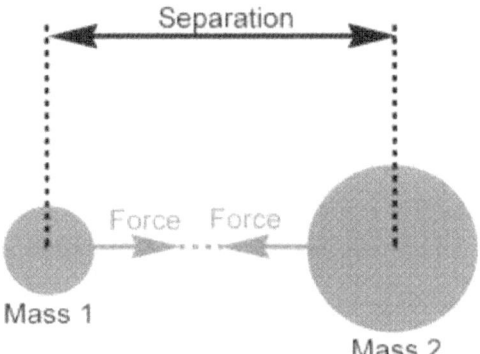

Figure 7: Force of attraction, is directly proportional to the product of their masses (M1 × M2) and inversely proportional to square of separation of bodies.

NEWTON'S THREE LAWS OF MOTION

1. *Newton's first law states that, "if a body is at rest or moving at a constant speed in a straight line, it will remain at rest or keep moving in a straight line at constant speed unless it is acted upon by a force". This postulate is known as the **law of inertia.***

2. *The second law states that the acceleration of an object is dependent upon two variables – "the net force acting upon the object and the mass of the object".*

3. *Newton's third law is: "For every action, there is an equal and opposite reaction."*

With the help of the gravitational field equations[10], he computed 'v', the velocity of motion of a planet in gravitational field of Sun and in turn, with these equations he could calculate the mass 'M' of Sun and of all other planets.

By noticing the 'fluctuations in the speed of motion of a planet in its orbit' he could even predict, far off existence of some undiscovered planets. This understanding of universe later on changed our lives in many ways.

But, Mr. Newton could not explain **'how' and 'why' this attraction occurs.** Many scientists observe only 'What', without exploring this 'how' and 'why' of a phenomenon, therefore these discoveries delivers incomplete picture as in the above story of Elephant.

Newton could not tell whether the 'gravitation is a long–range interaction' and as such no medium is required for this interaction, or it is a 'short range interaction' which requires a medium to propagate the force.

(This was a period when Ghosts and witches with their super natural powers and with their elusive appearances were a common belief. Newton himself has written some books on these subjects).

---

10   *Newton's equations describe the gravitational field by two field equations ($MG = rv^2$ or $MG = r^2a$).*

After **Newton's description of gravitational field,** *in 1915* **Einstein re–discovered the field equations of 'Gravity' in his 'general theory of relativity'.**

Mr. Albert Einstein (1879–1955)

In 'general theory of relativity' he could extend his 'theory of relativity' and 'equivalence principal' to study the gravitational field and **rightly** *calculated the configuration and extent of deformation of Space – time produced by Sun, as 'Schwarzschild matrix' its gravitational field, and 'Schwarzschild sphere' as gravitational sphere of Sun around its gravitational center.*

Figure 8: Einstein's deformation of space-time by bodies, and they move towards each other according to the difference in slope of deformation.

He observed that **'Planets move straight but the path is curved'** *as the Sun distorts Space–time around its surroundings, which curves the path of planets. His general theory of relativity is a 'short range theory' but his description of space–time distortion around a body as such never felt the necessity of a medium.*

Higgs field

Mr. **Peter Higg**s *proposed 'universal Higgs fields' which penetrates everywhere in universe.*

His description that when a particle moves in this omnipresent 'non zero energy field' the smaller particles e.g. photon passes unnoticed, while bigger particle electron feels resistance and all elementary particles drags the field along with them; and thus he advocated an omnipresent field which reacts to elementary particles.

The 'universal, uniform, multilayered field was independently proposed as Space Matrix (space-Time– Energy) by the author in 2005 in his 1st book 'Inside a wave'. It was earlier described as Higgs fields but its properties and working was never discussed before conception of this model.

For the familiar fields, such as the electromagnetic fields, the lowest energy state is the one in which the field have zero value (that is, the field vanishes), and if any non-zero field is introduced, the energy stored in the fields increases the net energy of the system.

But for the Higgs field, the energy of the universe is lower, if the field is not zero, but instead has a constant nonzero value i.e. in its natural, lowest energy state, the universe is permeated throughout by a nonzero Higgs field.

The Higgs field is a quantum field, but the fact is that all elementary particles arise as quanta of a corresponding quantum field. Particles that interact with the Higgs field behave as if they have mass, which is proportional to the 'strength of the field times the strength of the interaction'. **It can be understood as if an elementary particle is dragging some amount of Higgs field along with its motion by 'Higgs mechanism'.**

It is described in this Treo model, that dragged Higgs fields are bound treos from space matrix; (either as full kinetic coloumns, few layers of kinetic coloumns or in the form of localization energy as composite mass energy of unit electron).

In second dimensional deformation, these dragged bound treos contribute as increased **number of layers in all Kinetic coloumns of receiveing particle with** its increased momentum, or as increased localization energy which contributes in formation of kinetic coloumns of atomic (glueons) and weak forces (W,W+ and Z particles). (Table 1, page 47).

Without the Higgs mechanism, the W and Z bosons that mediate the weak force would be massless, just like the photon (to which they are related to), and the weak interaction would be as strong as the electromagnetic one.

*Thus, we have understood the three ways that mass arises: The main form of mass we are familiar with – that of protons and neutrons and therefore of atoms – **come from the motion of quarks** bound into protons and neutrons.*

The proton mass would be about what it is even without the Higgs field. The masses of the quarks themselves, however, and also the mass of the electron, are entirely caused by the Higgs field. Those masses would vanish without the Higgs field.

Mr. Nimbu and strings

The 'super string theory' proposed by Michael Green and John Henry Schwarz in 1984, however, assumes that the ultimate building blocks of nature consist of tiny vibrating strings. Fundamental forces and many particles found in nature are nothing more than different modes of vibrating strings.

Broadly speaking there are five string theories which can be combined in one M theory (Type I, Type II A, Type II B, Heterotic – o and Heterotic – E). These mathematical models mostly rely on symmetries. The study of symmetry is called 'group theory'[11].

---

11   The founder of this new branch of mathematics "group theory"[Ref 3] great French mathematician Evariste Galois was born in 1811. Using the power of symmetry alone. Galois as a teenager solved the problem of 'quantic (five powers) equation' which stumped the world's greatest mathematicians for five hundred years. Galois was far ahead of his time, and other mathematicians did not appreciate

All 'five string theories' and 'super gravity theory' is just different approximation of one 11–dimensional M theory. M theory allows 'curl up

extra dimension' in internal space only in a prescribed way. According to this M theory the strings are just one member of wide class of objects which can be extended in more than one dimension i.e. A 'p brane' has length in p directions.

The gravitational interactions, for example is caused by the lowest vibratory mode of a circular string (a loop). Higher excitations of the string create different forms of matter. At last, the 'Strings' serve as the basic object, in the direction to solve the problem and indicates that the gravitational force can also be combined with other three forces.

Very long 'cosmic strings', are also predicted by the theory. The 'filaments' of billions of light years long have been picturized which bind together 'multiple galaxies' or group of 'galactic clusters' and are the binding agent of whole universe.

While the speculated filaments, can be the 'graviton coloumns' on outer most layer of gravitational spheres of 'Cosmic bodies' and 'black holes', which join together to form gravitational fields of these bodies, (e.g. in the case of Sun, all graviton coloumns arising from periphery of its gravitational sphere combine side by side, and produces solar gravitational field, as proposed in present model). Thus, as such it allows putting 'elementary particle orbits' in second dimensional deformation with 'planetary orbits', in the same frame.

---

his path breaking research. His three submissions were ignored or dismissed as 'incomprehensible'. He died in 1832 in a *duel* with pistols.

Fortunately evening before the duel, Galois had a premonition of his death. He wrote down his key results to his friend Auguste Chevalier and desired it to be published in *'revue eyclopedique'. This was not published for 14 years. A century later, the mathematicians were still puzzled over his notes, because he made references to mathematical equations that were not discovered until 25 years after his death.*

The 'Strings' as described in all 'string theory' and 'super gravity theories' describe the properties of newly proposed 'energy particle', which has been named as 'treo'. The mathematical models of strings, super string and super gravity theories and M theory although describe some properties of 'bound treos' and 'unit space matrix'; but again, fail to deliver a model.

P1 brane is a string (can be seen here as one '**bound treo**') *and p2 brane is a two dimensional membrane* (which is described as **sheet** in two dimensional deformation of space matrix) *P3 is a three dimensional blob* (named as '**electron black hole**' produced by three dimensional deformation, in treo model) *but higher values of p are also possible up to 9, in ten or eleven dimensional space time.*

3. Photons

All Photons are packets of integral multiple of one quanta mass energy (S number of free treos) which increases at each of $\sqrt{S}$ quantum levels to form all $\sqrt{S}$ type of photon packets. These photons can be divided according to its uses and properties, as shown in table below and form EM spectrum.

In all photon packets their $E = h\nu$; *or E (energy mass) = h (planck constant)* × ν *(frequency of wave)*[12]. *The 1 frequency energy mass photon is hv1, and similarly hv2 is packet with frequency 2, and so on.... hv3, hv4, hvn (any n number) up to biggest gamma photon which is of hv$\sqrt{S}$ energy mass.*

In above relation *(E = hν)* if wave length of any photon packet is equal to the speed of light i.e. wave length of unit photon is S bound treo distance (distance travelled by light in one second) then its frequency ν will be 1. Now energy mass of photon packet of frequency 1 will be

---

12 Frequency is equal to number of quanta mass energy in any wave packet (This is expressed in proposed formula of wave length = S number of bound treos/number of quanta in photon packet).

equal to value of planck constant E=*h* .The S number of free treos or **1.85485844 × 10⁴³ free treos** or **one quantum energy mass**; is value of **Reduced Planck constant *ħ* (*h bar*). and angular momentum of this one quanta mass energy calculates the value of Planck constant 'h'.** (*ħ × 2π = h*) page 70.

Though photon have no rest mass, they do have electro magnetic energy [Ref 4]. This energy mass increase by one quantum at each of √S quantum levels of first dimension to form all √S types of photon packets and depending upon number of quanta mass energy in any photon packet, its frequency and wave length is decided.

All √S type of photon packets which form at √S quantum levels can be divided according to its uses and properties as shown in table below and form EM spectrum. Have we ever tried to know that **how** these photons deform space matrix, and how are they supported and propelled by it?

With increasing mass energy, all √S types of photons of EM spectrum spread on successively reducing 'wave length', which can be calculated according to proposed formula; wave length in bound treos = S bound treo length/number of quanta mass energy in this photon packet'.

This distributed load of any number of free treos exerted on each apex bound treo in its wave length, is neutralized **@ one free treo by one kineton.**

Moving photon packet can form **'Secondary wave let'** from any apex bound treo in its wave length, from where total mass energy of this photon packet can form one new wave front in new direction.

Increase of mass–energy by one quantum in photon packet at each of √S quantum levels in first dimension, also increases the angular momentum of each new photon packet by one more unit, and thus it forms one more EM wave per second i.e. one wave per quanta mass energy of photon packet. (Fig. 15 page 93)

Region	Wave length (Armstrong)	Frequency (Hz)	Free treos in packet in terms of S (= 1.855 × 10^{43})
Radio	> 10^9	< 3 × 10^9	< 3 × 10^9 × S
Micro Wave	10^9 – 10^6	3 × 10^9 – 3 × 10^{12}	3 × 10^{12} × S
Infra red	10^6 – 7000	3 × 10^{12} – 3 × 10^{14}	4.3 × 10^{14} × S
Visible Region	7000 – 4000	3 × 10^{14} – 3 × 10^{14}	7.5 × 10^{14} × S
Ultra Violet	4000 – 10	3 × 10^{14} – 3 × 10^{17}	3 × 10^{17} × S
X-rays	10 – 0.1	3 × 10^{17} – 3 × 10^{19}	3 × 10^{19} × S
Gamma Rays	< 0.1	> 3 × 10^{19}	> 3 × 10^{19} × S

Table 2: Mass energy Spectrum of EM radiation

Biggest gamma photon has √S quanta mass energy, which spreads on its √S bound treo wave length and thus exerts a uniform **load of one quantum (S free treos)** on each apex bound treo. This load of S free treos is neutralized by **equal** S number of kinetons (**as one kineton support one free treo)** in one √S layered sub kinetic coloumn, which form on each of √S apex bound treos in its wave length.

These √S layered sub kinetic coloumns (each having S kinetons) will form at last √Sth quantum level of first dimension, where this last new √Sth layer will have 2√S − 1 kinetons (as there are 2n-1 kinetons are in each nth layer). 2√S − 1 rows of √S length will be required to form last layer in all √S sub kinetic coloumns at all √S apex bound treos (in its wave length).

This means **two small squares (each having √S bound treos in its sides),** will get deformed to form these 2√S rows each of √S bound treos length.

As this packet will move on √S new apex bound treos in one line by √S vibrations to complete its one wave on √S wave length.

All photon packets slide in translational motion, from one bound treo to next bound treo of space–matrix with each vibration of space matrix i.e. at the speed of light.

All rotating sub kinetic coloumns together construct one transverse EM wave, and field of EM forces. Having no charge, many photons can pile up together to form photon beam and it make the EM forces very strong?

The number of EM waves formed per second = number of quanta mass energy in this photon packet = number of kineton layers in each one sub kinetic coloumn = frequency of wave = angular momentum of this photon packet.

4. Reduced Planck Constant, Planck Constant

S numbers of treos form one quantum energy. **The S free treos or one quantum energy** (which is mass–energy of a unit photon) **is the value of Reduced Planck constant $\hbar$ Angular momentum of this one quantum mass–energy is the value of Planck constant $h = \hbar \times (2\pi)$.**

The one quanta energy (*value of Reduced Planck constant $\hbar$*) produces unit action and deforms just one bound treo layer of unit space matrix and its one unit angular momentum (*value of h; Planck constant*) produces one EM wave in one second. (Fig. 15, page 93)

a. **Reduced Planck constant is Energy of one quantum mass which** is value of 'Quanta of unit action' **or** 'unit minimum action'= **1.0545718 × 10⁻³⁴ Joule sec.** (*conventional Codata 2018 value of Reduced Planck constant*)

b. *Planck constant is the* **angular momentum of this one quantum mass–energy'** *which produces one EM wave in one second.*

6.626070 × 10⁻³⁴ Joule sec. (*Conventional Codata 2018 value of Planck constant*)

1. *If we divide conventional value of Planck constant by 2π*

   6.626070 × 10⁻³⁴ Joule/6.28318531 (value of 2π)

   1.05457179 × 10⁻³⁴ Joule

   (we get conventional value of Reduced Planck constant per second)

2. **If we calculate Mass of this energy (m = E /c²).**

   $1.05457179 \times 10^{-34}$ Joule/$(2.99792458 \times 10^{8}$ meter per second$)^2$

   $1.17336936 \times 10^{-51}$ kg (mass of one quanta energy)

3. **Number of free treos in this mass of one quanta mass**

   1.173369×10^{-51} kg × 1.5808×10^{94} (number of free treos in one Kg) *

   = **1.855×10^{43} treos (S number of free treos)**

   (S free treos = one quantum energy, which is present in unit photon)

   *see calculations for new derived value of gravitational constant page 218.

5. 'Dualistic Nature' of Photon and All Matter Particles

All Photon, electrons and all elementary particles sometimes presents themselves as a wave and sometimes as a particle, and thus they all exhibit wave particle duality.

The photon packet can move at the speed of light, **only after** distributing its mass energy on its supporting kinetic coloumns @ **one free treo on one kineton,** as they are always in equal number. Thus, all free treos as mass energy of any photon, gets distributed on all Kinetons present in all sub kinetic coloumns in its wave length and **thus after concealing its mass energy**, it moves as **'photon wave' at the speed of light**. (Fig. 9 b)

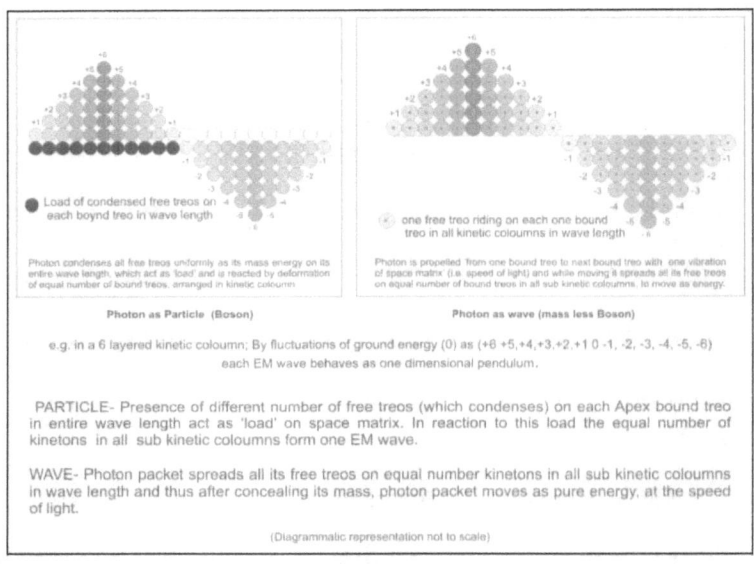

Figure 9 a and 9 b: Wave particle duality of photons and electrons & all elementary particles.

Any type of packet **(when confined) distributes its total mass energy uniformly on all apex bound treos along its RC wave length and thus behaves as particle**. (Fig 9 a)

All type of photon packet while in continuous wave motion, in fact **conceal mass energy of all their free treos by spreading them on equal number of kinetons** present in all supporting kinetic coloumns, (which also explains 'no rest mass of photon). (Fig. 9 b)

We started to utilize this property of space matrix, when we started to send photons and received them as radio waves in our radios. We extended this use of free treos when we started to send whole frames of one–dimensional deformation of local matrix by bunch of photons of different frequency simultaneously and could receive them on our

Telivision sets. We mastered to deform local space matrix by photon (mass) energy to form wi –fi zones by which we use internet, send e–mails and Whatsapp etc.

This masking of mass energy and its spread on supporting kinetic coloumns @ one free treo on one supporting kineton, explains **mechanism of double slit experiment, quantum tunneling and execution of weak forces by W, W+ and Z bosons to cross nuclear and atomic barrier while transporting heavy alpha and beta particles.** (Fig 9 b) In this way, the 'wave particle duality' is exhibited by all photons and elementary particles.

But when photon packet is confined, then it behaves as a particle and one energetic photon of sufficient mass energy (frequency), pushes free electron on stricking a metal surfaces to produce **photo electric effect** and thus it generates electric power. (Fig. 9 a)

Similarly, in any electron packet (or in any elementary particle) each quanta mass energy (S free treos) is supported, as it spreads on one orbitum (each made up of equal number of S kinetons), and thus total number of quanta mass energy in any such packet spreads on equal number of orbitums in all sub shells and shells forming its matter wave. (Fig 9 b)

Or it can **condense and uniformly distribute this electron mass energy of free treos, on all apex bound treos along its RC wave length** and this load on each apex bound treo is supported, rotated and propelled by each one shell present at each apex bound treo in its vertical wave length, and then this electron behaves as a particle. (Fig 9 a)

6. Pair Production

One gamma photon packet of 1.02 MeV (**√S quanta mass energy**) will exert **one quantum load of S bound treo**s at each of √S apex bound treos in its √S wave length; which will be neutralized by equal number of S kinetons present in one √S layered sub kinetic coloumns formed on these each apex bound treos.

Due to kinetic push by these S kinetons in all sub kinetic coloumns (equal to ground energy of first dimension) the supported gamma photon, **instead of being shifted linearly to next bound treo with next vibration, will start rotating at same apex bound treos.**

Due to this rotation of mass energy, each one quanta energy of S free treos at each apex bound treo will be converted in S **treo charge;** where its **anticlockwise rotations gives it negative charge** (as in electron) and **clockwise rotations (as in positron) gives positive charge. (page 158)**

The supporting kinetic coloumns at each apex bound will also rotate, and vibrate S times (as a unit) to support this one quanta rotating load, from S planes (in 360°) by S vibrations in one second, while forming one EM wave.

Each load will remain stationary at its individual apex bound treo (but rotate) without advancing forward, **till √S vibrations occur to support this packet,** (one vibration of each of √S coloumns in its wave length); and only **after this by one next propelling vibration this packet will shift** to next newly **formed apex bound treo in its newly formed orbit,** and this process will be repeated √S times in one second (in S vibration) to move this packet by √S bound treos distance in one second in its orbit.

The anti clockwise revolution in orbit is of the electron (**particle**), while positron will revolute clock wise in its orbit to make it **anti particle.**

This gamma photon energy in its wave motion will divide to form one pair of electron and positron, **by bifurcation of each sub kinetic coloumn,** as two rails on railway track curles in two circles on both sides.

and thus out of 2 √S rows newly deformed at last quantum level, √S rows will be used by one electron and √S rows will be used by positron, which will form √S orbitums for each of two packet.

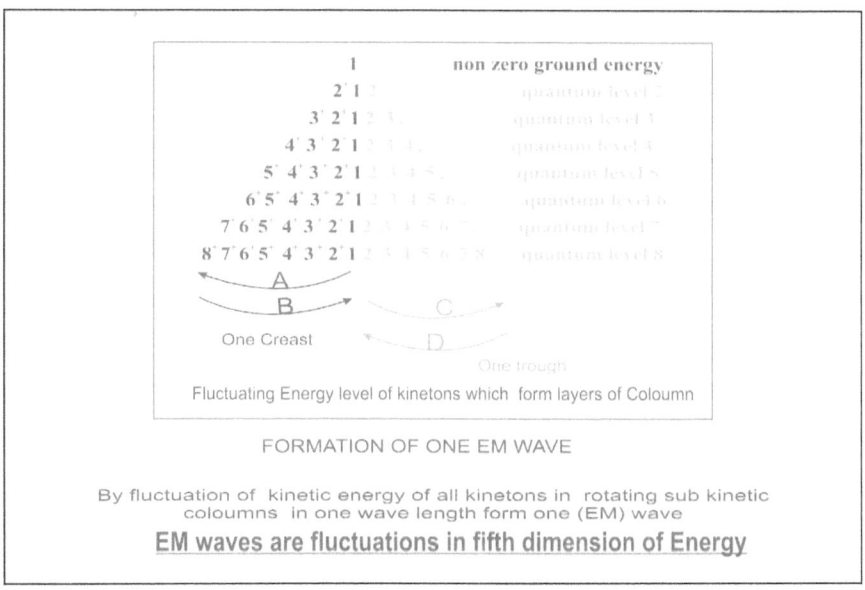

Figure 10: Energy fluctuations along its one EM wave.

√S length of each row will form √S radius of one orbitum, and while rotating it will make a disc of π (√S)² area and of 2π√S circumferance.

Thus the orbitums are formed at each apex bound treo (which replaces kinetons in sub kinetic coloumns). **Each orbitum formed will now support one quanta load by its one rotation in one second.**

One EM wave is Pendulum motion of one quanta mass-energy, as number of waves formed are equal to total quanta mass energy in any packet; to support this packet in one second.

Breaking of EM wave of Gamma photon
To form Electron- Positron Pair

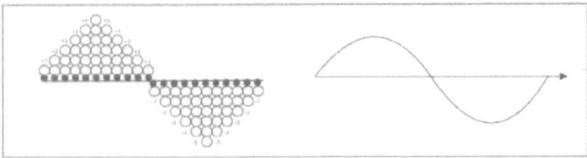

Increasing quanta & angular momentum in photon packet increases height (frequency) while its length (wave length) contracts. The gamma photon packet breaks as two lobes (electron and positron), one from 'creast' rotating clockwise and other from 'trough' rotating 'anti clockwise'

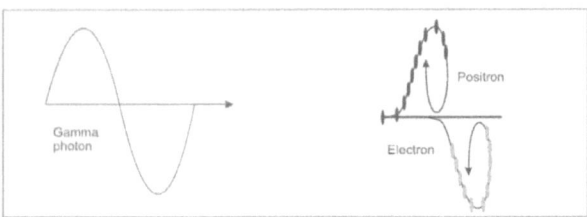

Figure 11: Breaking of Gamma Photon EM wave for pair production: EM wave of gamma photon which divides 2√S rows, where trough part start rotating anti–clockwise to form unit electron, while mass energy on crest part in another half of EM wave on √S bound treos will rotate clockwise to form a positron.

Sine waves are pendulum fluctuations of kinetic energy of all 2n-1 kinetons in any one layer of kinetic coloumn e.g. 15 kinetons present in outer most 8th layer in one 8 layered kinetic coloumn, fluctuates its energy from base energy between + 8 to – 8, as in a pendulum varying from ......-8, – 7, – 6, – 5, – 4, – 3, – 2, – 1, 0, +1, +2, +3, +4, +5, +6, +7, +8 ... and also provides different magnetic momentum to each kineton in this layer (as in Fig 10).

After reaching the maximum limit of all factors which decides the formation of EM sine wave, the wave will divide. **These limits are**

(1) when the load supported at each apex bound treo reaches the maximum limit of **one quanta** or S free treos, **equal to the base energy of first dimension.**

(2) when increasing **angular momentum reaches to maximum of √S units** which form √S EM waves in one second. As after this increase

of angular momentum by one more unit the wave will get converted into a circle.

(3) when fluctuation of base energy in 2 √S rows (which form last one biggest layer of all √S sub kinetic coloumns) +√S energy which will form one positron – √S energy which will form one electron. (Fig 10 & Fig 29 on page 110)

(4) Finally at this last √S quantum level, **the length of packet** on √S bound treos (wave length) **becomes equal to its breadth** with maximum √S frequency for EM wave (√S layers in kinetic coloumn) for this √S quanta mass energy in packet. (**NO EM WAVE CAN FORM; with frequency more then its wave length**)

Any further increase of mass energy will make this gamma photon of 1.02 MeV unstable and it will break in two, both packets will **start revolving in its individual orbit;** anti–clockwise to form unit electron of 0.511 MeV i.e. √S quanta mass energy, while mass energy in another half will revolve clockwise to form a positron of 0.511 MeV mass energy or √S quanta mass energy. This process of breaking of energetic gamma photon is called 'period doubling bifurcation' (Fig. 11)

Angular momentum divides in spin angular momentum and orbital angular momentum, now from spin 1 in photon packets it becomes spin ½ for electron packet; i.e. **particle will regain its origional orientation only after two rotations of packet.** By one vibration of all its shells on each of √S apex bound treo in its RC wave length, one by one from top to bottom by which packet will rotate once and then back again from bottom to top this packet will rotate once more, to regain its origional orientation thus forming its **one matter wave.**

CHAPTER 3

Deformation of Space-Time

1. Deformation of Four Dimensions of Space-Time

We will study in detail the geometry and step by step construction of fields of all fundamental forces by increasing deformation in all four dimensions –

Universe has only one force, which is generated by kinetons in any type of kinetic coloumn (small or big) and is produced by **Action-reaction mechanism of space matrix.**

For ACTION, initiated by applied load of any mass or momentum ranging from **one free treo to S quanta mass energy** (S x S free treos) of unit mass, the space matrix REACTS, by converting equal number of bound treos of space matrix in kinetons and forms one reacting kinetic coloumn to neutralize this action by producing an equal and opposite reaction.

But the reacting kinetic coloumn changes its geometry and amount of force with inclusion of increasing number of dimensions of space–time, then the nature of force also changes; it changes from EM forces to atomic and weak forces and then gravitational fields and gravitational spheres are formed.

In our **quantum universe** the mass energy does not increase linearly, but it increases in steps, at each of $\sqrt{S}$ **quantum levels which are formed in each of four dimensions of space–time,** and is supported by addition of one new layer at this quantum level, which is added in this kinetic coloumn and then equal number of kinetons in kinetic coloumns act togather to neutralize the load of this mass energy.

This geometry of deformation of unit space matrix changes, along with the nature of force generated, as the deformation increases (with increasing load) and involves all four dimensions one by one. Thus with increasing load it, successively involves one (length), two (length and breadth), three (length breadth and depth of unit space matrix) and then all four dimensions of space–time.

With increase of **one quantum mass energy of one unit photon** at each of √S quantum levels in **first dimension**, it **produces √S type of all photon packets of EM spectrum.**

The photon packet spread uniformally on all apex bound treos along its (gradually reducing) wave length and exert equal load, on each of its apex bound treo in wave length. (where wave length = S number of bound treos/number of quanta in packet)

To support this load equal number of kinetons are generated from bound treos of space matrix to accomplish **Action-reaction mechanism,** and get arranged in **transverse sub-kinetic coloumns of 1 to √S layers,** at each apex bound treo along wave length and thus the **EM fields** of EM forces are generated.

As the deformation increases, with increase of √S **quanta mass energy (of one-unit electron)** at each of √S quantum levels of **second dimension,** it supports all elementary particles, nucleons, atoms, molecules and masses **up to one unit mass** (*Planck mass: 2.176 43 × 10⁻⁸ Kg*) **or S quanta mass.**

With increasing mass energy on its reduced wave length the load on each apex bound treo increases at each nth next quantum level by **square of quanta** i.e. n^2 **quanta.**

To support its each quanta load (S free treos) one orbitum (of equal number) of S kinetons is formed, by Action-reaction mechanism at each apex bound treo along its RC wave length.

The total number of orbitums are always equal to the mass energy quanta in any elementary particle packet, which are formed in sub shells of shells present at each apex bound treo along its RC wave length; thus at matching quantum level all elementary particles and their matter waves are formed. Finally with completion of this √S layered full kinetic unit gravitational coloumn which is now named as graviton coloumn of second dimension, the biggest mass supported by second dimensional deformation is unit mass (the mass roughly equal to size of one flea egg).

One unit mass is the maximum load, which can be supported at one graviton, (which is the apex bound treo of one graviton coloumn now named as Graviton) forms at its unit gravitational centre.

The body made up of multiple unit masses and its load (in square number of unit masses in body exerted at its unit gravitational centre) is supported by equal number gravitons arranged in kinetic coloumn of third dimension i.e. 'one electron black hole'.

By increase of one-unit mass (or S quanta mass energy) at each of √S quantum levels in third dimension, finally one unit space matrix (its S sheets) by its full deformation, produces √S number of spiral layered 'electron black hole' (the kinetic coloumn; having S number of gravitons) which can support √S unit masses (approximately 10^{13} Kg or one billion metric ton), and its load of S unit masses (in square of unit masses in body) at its unit gravitational center.

Still bigger (any) cosmic body, of n unit masses exert a load of n^2 unit masses at its gravitational centre which is supported by n^2 numbers of gravitons in this gravitational sphere, with the deformation of all four dimensions of space–time.

After the deformation of all three dimensions of space or full deformation of one unit space matrix, which form one electron black hole, the deformation of SPACE-TIME starts at √S **quantum levels of fourth dimension.**

In formation of one **gravitational sphere** (kinetic coloumn of fourth dimension) each next n^{th} quantum level is formed by addition of one new layer (the nth layer) having 2n-1 electron black holes (2n-1 in any n^{th} layer and total n^2 in n layered coloumn according to coloumn geometry), to support multiple unit mass body at its gravitational centre.

With increase of √S unit masses (as one unit) at each of √S quantum levels **in deformation of fourth dimension of space time**, finally at last quantum level by formation of S bound treo layered gravitational sphere (√S bound treo layers at each quantum level x √S quantum levels); **S unit masses (10^{43} unit masses) of one unit black hole will be supported at its gravitational centre.**

The **Sun** made up of 10^{38} unit masses as its mass energy, is supported by 10^{38} **bound treo layered[13] gravitational sphere** around its gravitational center, while our **Earth** made up of 10^{32} **unit masses** have 10^{32} **bound treo layers** (in 1 mm) **in its gravitational sphere.**

At periphery of any gravitational sphere of any **n unit mass** cosmic body, there are 2n-1 gravitons in its outer most n^{th} layer. **The 2n-1 graviton coloumns (one at each graviton) at outer most layer of this gravitational sphere merge together side by side to form gravitational (coloumn) field of this body.**

---

13 Same size gravitational sphere of 3 Km diameter ($2x\ 10^{38}$ bound treo layers) of sun is calculated by Einstein in his general theory of relativity.

At gravitational centre of any cosmic body, at which it support the central M^2 load of M unit mass body[14] (supported from all directions and in all plains), will exert 2MG load (in one direction and in one plain), on the outer most layer of its gravitational sphere and also at any layer in gravitational field of body.

This 2 MG load, is supported by 2n-1 gravitons at outer most layer of gravitational sphere; or by 2n-1 kinetic coloumns which are formed at each apex bound treo in its RC wave length of any orbit (layer), in two dimensional deformation of gravitational field.

Sum of kinetic energy of $v^2 \times$ (2n-1) kinetons, in all 2n-1 kinetic coloumns in one matter wave (on 2n-1 apex bound treos in any n^{th} layer of gravitational field) supports this 2 MG load of body, and **thus each concentric circular layer of gravitational field by its one wave can individually support the total load of body (sun), exerted on it.**

The **gravitational field (coloumn) gradually dilutes** first in spherical three dimensional deformation (in which mass of body is accommodated e.g. spherical body of Sun), then on two dimensional deformation (in which orbits of all baby bodies are forms at 10^{4th} quantum level) and finally it dilutes in one dimensional deformation, which extends till last bound treo layer of this gravitational field, where finally one free treo load is supported by one kineton.

2. Unit Space Matrix

This S number is COSMIC CODE which governs and carves this QUANTUM PENDULUM UNIVERSE.

Universe is vibrating S times per second at COSMIC RHYTHM. Space matrix is omnipresent and the presence of load on it at any point,

---

14 (M is mass of body in Kg and G is Newton's universal gravitational constant; **when both multiply it calculates the, number of free treos which forms in M unit mass body**)

mark the boundaries of one unit space matrix. Starting from this point the cosmic rhythm, in presence of load, **by its S vibrations quantities space-time-energy; S number** of vibrations simultaneously demarcate; **unit space, unit time** and **unit energy**; to quantities all five dimensions of Space-matrix.

UNIT SPACE, is **a cube with its each side of S number of bound treos,** (the distance traveled by light in one second), decides **one unit space.** S number of vibrations which occur in time period of S number of Planck least times mark **one** UNIT TIME **of one second.**

The quanta of UNIT ACTION PER SECOND consume S vibrations, S number of bound treos (S kinetons) is one quantum kinetic energy which is required for ONE UNIT ACTION, and is calibrated by Reduced Planck constant (see calculations, on page 70).

A cube having S number of bound treos on its one side is ONE UNIT SPACE-MATRIX **(UNIT SPACE – UNIT TIME – UNIT ENERGY); governed by this rhythm of universe, of S vibrations per second.**

One quanta mass energy or S number of free treos (in one–unit photon) evenly spreads **on one side of this cube on S number of bound treos,** where one free treo load on one (apex) bound treo is supported by one kineton, by its S **vibrations which occur in one second,** (or by one vibration it is supported for the period of one Planck's least time and along with it is pushed to next bound treo in direction of motion; to travel at the speed of light).

This quantified one–unit space systematically contract by increasing load at this point, along with involving increasing number of dimensions of space –time, to produce fields of all four basic forces (EM force, weak and atomic forces and gravitational force).

Unit space matrix is a cube with all its sides having S bound treos, can be better understood by simile of a book, that there is **one line of S bound treos, S such lines in one page** (S^2 bound treos), **and S such pages in one book** (S^3 bound treos).

S vibrations which can involve only S number of bound treos in one dimensional deformation of length; with changing coloumn geometry in increasing number of dimensions, will effect S^2 bound treos in two dimensional and S^3 bound treos in three dimensional deformation of unit space matrix, by same S number of vibrations.

Our universe is quantum universe and our space is also quantified by S, S^2 and S^3 bound treos in one, two and three dimensions, which together form one cube of one unit space matrix, and it vibrates as one unit continuously by S vibrations (per second) per unit time.

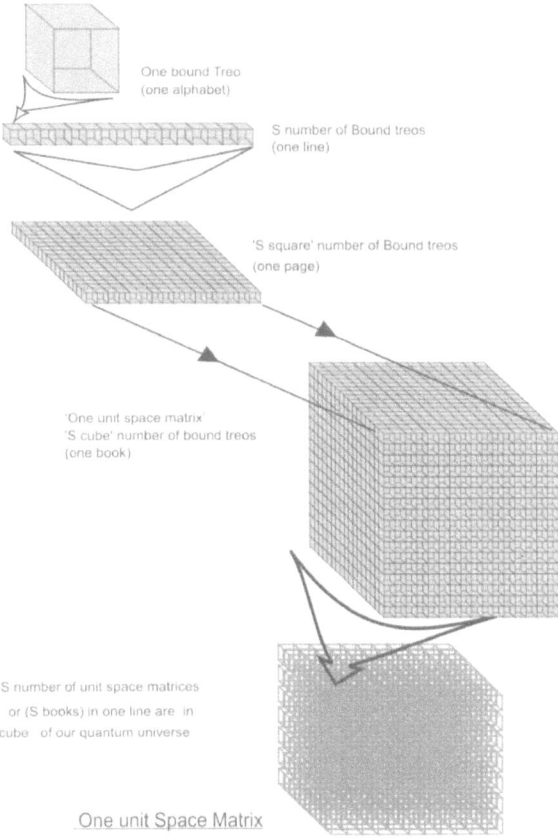

Figure 12: One Unit space matrix a cube with S bound treos present on its one side

With **2 S number of such unit space matrices** on each side of the biggest cube of one–unit quantum universe; the **functional universe**[15] will have **S space matrixes (S² bound treos) in its radius** (as postulated/calculated).

3. Coloumn Geometry

One kinetic coloumn forms at **each apex bound treo** along wave length of photon packet or RC wave length of elementary particle packet to support its exerted 'load'. **Both Linear deformation** and **angular radial deformation** occur simultaneously, but for sake of description they are described here separately.

1. Linear Deformation

The photon packet and all elementary particles packet spreads on space matrix along a line, and with its uniform packet density it exerts uniform equal 'load' on each (apex) bound treo present along its wave length.

This 'load' (mass/momentum) is perceived as an action by each apex bound treo & reacted by formation of one 'kinetic Coloumn' from local space matrix. Kinetic coloumn made up of equal number of kinetons are formed to support this 'load' at each apex bound treo, **as each one free treo of 'load' is supported by one kineton.**

(a) Kinetic Coloumn in deformation of first dimension

In deformation of first dimension, with increasing mass energy by **one quantum at each next quantum level will form bigger photon packets,** and at any n^{th} quantum level, the photon packet of n quanta mass energy

---

15 [If any part of universe still exist even beyond this limit, **it is a lost universe for observer**; as no signal coming from outside this zone and trying to enter in, at the speed of light, **will only hover** at this point (as it will not be able to penetrate in and vice versa any signal will also not go out of universe) **as this border of universe is itself racing out (swelling) at the speed of light.**]

will spread in one line on each apex bound treos along its gradually reducing wave length (wave length = S bound treo/n quanta mass energy in photon packet) and thus with increasing mass and decreasing wave length exert a uniform load of n^2 free treos at each apex bound treo.

With increase of one more quantum mass energy which generates one new **photon packet of n quanta at next n^{th} quantum level, one new layer is added having (2n−1 kinetons),** while each 'n' layered kinetic coloumn (one on each apex bound treo along its wave length), **will now have n^2 kineton to support this n^2 free treos exerted load** (square of nth quantum level of first dimension) **of n quanta mass energy packet.**

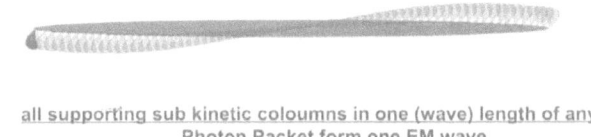

all supporting sub kinetic coloumns in one (wave) length of any
Photon Packet form one EM wave

Figure 13: Photon packet uniformly spreads on each apex bound treo in its wave length and all Sub − kinetic together form its one EM wave.

The total **n number of layers in 'kinetic coloumn'** of this **n quanta mass energy denote n frequency of wave** (or number of waves per second) **and also the n amplitude of wave.**

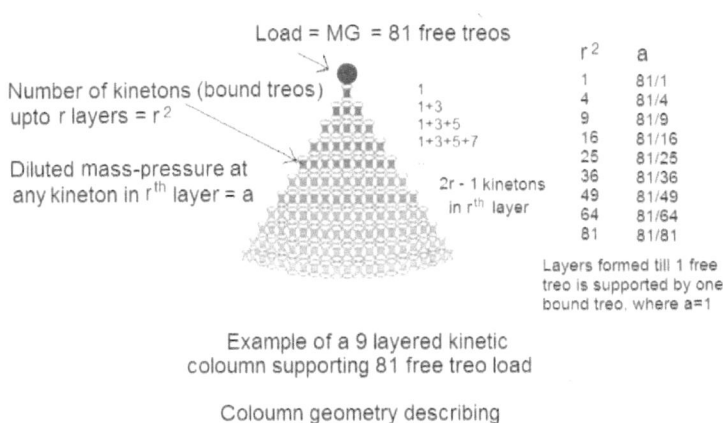

Load = MG = 81 free treos

Number of kinetons (bound treos) upto r layers = r^2

Diluted mass-pressure at any kineton in r^{th} layer = a

1
1+3
1+3+5
1+3+5+7

2r − 1 kinetons in r^{th} layer

r^2	a
1	81/1
4	81/4
9	81/9
16	81/16
25	81/25
36	81/36
49	81/49
64	81/64
81	81/81

Layers formed till 1 free treo is supported by one bound treo, where a=1

Example of a 9 layered kinetic coloumn supporting 81 free treo load

Coloumn geometry describing Newton's equation MG = r^2a

Figure 14: Coloumn geometry which is valid in all four dimensions

We will take the example of 9 quanta photon packet, causing a deformation at 9th quantum level of first dimension (see Fig. 14). To support the load of 9^2 free treos (81 free treos) at each apex bound treo, this one sub kinetic coloumn will have $(1+3+5+7+9+11+13+15+17 = 81$ kinetons) or 9^2 **kinetons** in this **9 layered sub kinetic coloumn.**

(b) Kinetic Coloumn in deformation of second dimension

In contrast to first dimensional deformation where mass energy was increasing by one quantum at each next quantum level; in deformation of second dimension, the 'mass energy or momentum in new packet' **increases by one 'unit electron mass' or by √S quanta (i.e. √S × S free treos)** at each of total √S quantum levels of second dimension.

One sub kinetic coloumn of first dimension is replaced by one rotating **Shell** in second dimension, and in place of one kineton layer it is now one **sub shell** (made up of √S bound treo layers), and in place of 2n-1 kinetons **2n-1 orbitums** are there.

Any 'n unit electron mass energy packet', of any elementary particle at n^{th} quantum level of second dimension, will spread on its **RC wave length** (S number of bound treos/number of quanta in photon packet).

It will now exert load of **n^2 quanta** free treos at each of these apex bound treos (square of nth quantum level of second dimension) thus each shell present on each apex bound treo along the R C wave length of matter wave at 'n^{th}' quantum level of second dimension, **will have n sub shells and n^2 orbitums; to support this load.**

In two (second) dimensional deformation at each next n^{th} quantum level, the additional 2n−1 quanta load is supported by addition of 2n−1 orbitums in one new n^{th} sub shell in each shell and each orbitum is made up of one quantum kinetic energy or S kinetons.

Thus at first four quantum levels it will form *one, three, five, seven* orbitums (2n-1 orbitums) with succesive addition of first, second, third and fourth sub shell, (Fig. 27, page 119).

Finally at last quantum level, $\sqrt{S}$ **unit electron mass energy or S quanta mass energy in 'one–unit mass'** is supported by (square number of quantum level) addition of $\sqrt{S}$ **sub shells having $2\sqrt{S}$-1 orbitums (2n-1) in this one shell** (as RC wave length reduces to just one bound treo).

Finally at last quantum level of second dimension the total deformations forms one biggest S layered (as $\sqrt{S}$ bound treo layers were increased at each of $\sqrt{S}$ quantum levels) **shell and $\sqrt{S}$ layered graviton coloumn,** which support one unit mass (one Planck mass i.e. 2.176×10^{-8} kg) with total deformation of second dimension at one 'unit gravitational center' or at one graviton.

In second dimension the **total number of bound treo layers** (new $\sqrt{S}$ bound treo layers are increased to form one new sub shell, at each next quantum level of second dimension) **in each shell decides frequency** of wave, which is equal to the **number of quanta mass energy** in this elementary particle packet.

(c) Three and four–dimensional deformation.

(In three dimensional deformation it is formation of one electron black hole, which can support the loads from one unit mass to one billion metric ton)

One unit mass is the maximum load which can be supported in universe at 'one bound treo or unit gravitational centre' by one graviton by its one graviton coloumn.

Therefore, **multiple unit mass body** or any cosmic body, **which exert square number of unit mass load at their gravitational center,** is supported by equal number of gravitons; firstly in **one electron black hole;** A kinetic coloumn of third dimensional deformation (for up to one billion metric ton) and then in **one gravitational sphere** of body a kinetic coloumn of fourth dimensional deformation, for still bigger cosmic body.

In short –

The coloumn geometry is always obeyed in the deformation of all four dimensions and as such, there are 2n−1 units which increase at each nth quantum level (total $\sqrt{S}$ quantum levels in each dimension) and n^2 units are in any n layered kinetic coloumn.

These 2n-1 units are **kineton in first dimension, orbitum (made up of S kinetons) in second, graviton (made up of S^2 kinetons) in third and finally 'electron black holes' (made up of S^3 kinetons)** in fourth dimensional deformational.

All these kinetic coloumns supporting load from one direction rotates with each vibration to support this load from all directions with time and thus they forms EM waves, matter waves, gravitational fields and gravitational spheres.

All thermodynamic, kinetic or chemical energy transfers are transfer of these full layers of kinetic coloumn from, the kinetic coloumns of donating body to the kinetic coloumns of receiving body.

2. Distribution of Load in Any One Kinetic Coloumn

Load of any number of free Treos, which is exerted **on each apex bound treo** along wave length, is supported by equal number of 'kinetons' which get arranged in the shape of one 'sub kinetic coloumn'. e.g. 81 'free Treos' 'load' is supported by equal 81 'kinetons' in this 9 layered one kinetic coloumn.

The total load exerted at apex of one sub kinetic coloumn distributes **equally on all kinetons, up to any layer in coloumn which is taken in account.**

If you take account of second layer, the load divides as ¼ on all three **kinetons** (diluted mass pressure 'a') of second layer (as total 1+ 3 = 4 kinetons), while **one kineton of first layer supports full load.**

When third layer of sub kinetic coloumn is taken in to account, it is $1/9^{th}$ of total load (diluted mass pressure 'a') **on each of 5 kinetons of third layer**, as there are total nine bound treos up to three layers in coloumn, (while it was ¼ of total load on each kineton in second layer, and full load on one kineton of first layer).

Similarly, it distributes its $1/16$ load (diluted mass pressure 'a') **on each of 7 kinetons in fourth layer** as there are total '16 bound treos to support the load up to fourth layer in coloumn', while it exerts $1/9$ of total load on each kineton of third layer (as it divides on nine kinetons) and ¼ of total load (as it divides on four kinetons) is exerted on each kineton of second layer.

But in this coloumn geometry this above mentioned load distribution pattern, describes **'a' as diluted mass pressure of body at each kineton in any layer of its kinetic coloumn.** This 'a' is the same as acceleration in Newton's gravitational field (thus this 'a' diluted mass pressure, can also be calculated by Newton's equation, $a = MG/r^2$).

All fields are made up of kinetic coloumns, and this coloumn geometry obeys 'inverse square law'. The gravitational field, intensity of light, charge density etc; all fade *'by reciprocal of the square of distance 'r'.*

So, it can be seen that **Newton's equations which describe gravitational field are formed in accordance with this proposed coloumn geometry.**

3. Quantum Gravitation in First and Second Dimension

One free treo load, is supported by one kineton, by its one vibration, for the time period of one vibration which occur in one planck's least time and total S vibrations in one second will support this free treo for a period one second (continuously).

Thus any quanta mass in packet will be supported by equal number of EM waves or orbitums in 'matter waves', and each such wave supporting one quanta mass energy will use S layers to form one EM waves; or one orbitum will support it by rotation of one layer of √S kinetons in S number of planes by S vibrations in one second.

Frequency of **any photon, electron or any elementary particle packet** is always equal to the **number of quanta mass energy in its packet,** or **number of EM waves formed in one second** or **equal to number of bound treo layers deformed** in kinetic coloumn up to this particular quantum level and also equal to **number of rotations of packet will perform in one second** and also equal to the **number of bound treo distance this packet will revolve in its orbit in second dimensional deformation.**

The load of S quanta mass energy of one unit mass is supported by total formed S orbitums in its one last layer and also in one graviton coloumn.

It also explains the **principal of equivalence,** as why number of free treos (which decides its **gravitational mass)** are always equal to kinetons (which decides its **kinetic mass)**

4. Angular Radial Deformation

With increasing energy mass by one quantum (value of reduced Planck constant) in each next bigger photon packets at each next quantum level in first dimension; one **more layer is added** in its all supporting sub kinetic coloumns (present at each apex bound treo in its entire wave length) and thus one layered to √S layered sub kinetic coloumns are formed.

With increase of this **one layer in all sub kinetic coloumns the angular momentum of packet and frequency of its wave also increases by one unit,** thus, Planck constant 'h' calibrates layer by layer deformation of space matrix in first dimension.

Similarly, in the second dimensional deformation the **frequency increases by √S number at each next quantum level,** due to addition of √S bound treo layers at each next n^{th} quantum level, (with the formation of one new 'sub shell' in each shell), along with increasing **angular momentum by √S units.** The radius of shell also increases by √S bound treo layers, at each nth quantum level; where n can be 1, 2, 3.... up to √S.

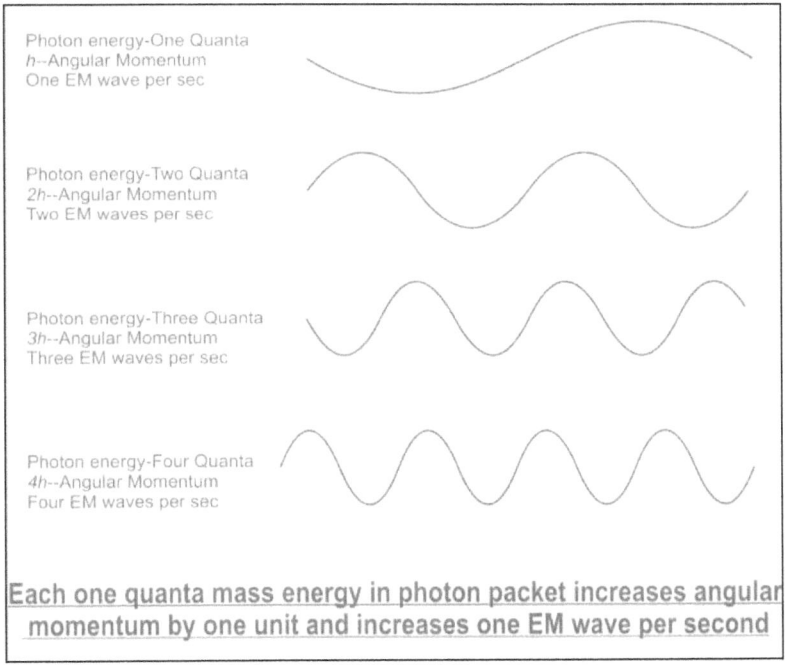

Photon energy-One Quanta
h--Angular Momentum
One EM wave per sec

Photon energy-Two Quanta
2h--Angular Momentum
Two EM waves per sec

Photon energy-Three Quanta
3h--Angular Momentum
Three EM waves per sec

Photon energy-Four Quanta
4h--Angular Momentum
Four EM waves per sec

Each one quanta mass energy in photon packet increases angular momentum by one unit and increases one EM wave per second

Figure 15: changing wave pattern with increasing angular momentum by one unit with increasing mass energy in photon packet by one quantum.

In case of 'Electron black hole' in three–dimensional deformation increasing angular momentum is decided by the increasing number of gravitons (and its wrapped graviton coloumns) **in each spiral layer (circles) of electron black hole.** The angular momentum increases as n (hC) in this kinetic coloumn, where n can be 1, 2, 3... up to √S.

While in deformation of four dimensions of space –time in 'gravitational sphere' of bodies, increasing angular momentum represents the **increasing surface area of gravitational sphere**, [which is calibrated with the **increasing radius of solid angle**, as n $(h/4 \pi)^{1/2}$, where n can be 1, 2, 3.... up to $\sqrt{S}$].

With this increasing angular momentum, the 'coloumn' turns from 0–90 degree, 90–180 degree, from 180–270 and finally from 270–360 degree, respectively in one, two, three and four–dimensional deformation of space matrix.

5. Proofs in Favor of This 'Coloumn Geometry'

1. "The **gravitational force, intensity of light, and charge density** all fade by *'reciprocal of the square of its distance'*, which indicates that the force fields are made according to this proposed 'coloumn geometry' (inversely proportional to square of distance) .

2. And there are direct proof with the distribution of electrons in *s, 3p, 5d, 7f* electron orbits in sub shells of shell which form **atomic structure**, is also according to this proposed coloumn geometry; of increase of 2n-1 units (1, 3, 5, 7) in n^{th} layer of proposed kinetic coloumn.

3. Solar gravitational fields

(!) **This 'coloumn geometry'** is also seen in the formation of 'gravitational field (coloumn) of a body' **where Newton's gravitational field equations (MG= r^2 a) describes the 'load distribution pattern in kinetic coloumn'** (page 90).

(!!) In the gravitational field of body, *gravitational kinetic energy, v^2* **decreases by *'reciprocal of its distance'*, is also in accordance with the coloumn geometry (v^2 = MG/r).**

In this equation v^2 is total number of kinetons in one v layered kinetic coloumn. According to coloumn geometry, it explains that this v is equal to the number of layers in each kinetic coloumn, and equal to frequency of wave which form in this orbit, and also equal to the bound treo distance per second by which baby body will move in its orbit.

(!!!) Any baby body will produce elliptical orbits in gravitational field, and thus motion of all planets is in 'elliptical orbits' as they move in a 'solar gravitational coloumn (field)'. It is again due to the **distribution of gravitational kinetic energy in gravitational field as per 'coloumn geometry'.**

(!V) The main baby bodies are placed between 10^{4th} to 10^{5th} quantum levels of gravitational field of parent body (e.g. satellites of 'outer planets', planets of solar system and stars in a galaxy are placed at 10^{4th} quantum level in the gravitational field of its parent body). The position of these baby bodies **and their speed of revolution in its orbits verifies the proposed 'coloumn geometry' and proposed model.**

4. According to **Schrödinger's equations for motion of a linear harmonic oscillator in second dimension** i.e. any moving load produces ripples at distance of 1, 3, 5, 7, 9, 11,13,15 units, **the same pattern decides, Quantum levels in gravitational field of any parent body.**

Thus Schrödinger's equations describes coloumn geometry (whose values are shown in the foot note[16]) and as the formation of orbits and atomic

---

16 Schrödinger's equations for motion of a linear harmonic oscillator

$Y = 1/2e-k^2/2$ when $a/b = 1$

$Y = 2Ke-k^2/2$ when $a/b = 3$

$Y = 4K^2-2)e-k^2/2$ when $a/b = 5$

$Y = 8K^2-12K)e-k^2/2$ when $a/b = 7$

it can be calculated and so on .... 9,11,13,15 ....

Proper values of a/b are given by $2n+1$ (n is an integer), the values of Y corresponds to characteristics eigen – functions while eigen value is wave length (length) of elementary particle.

structure is governed by these equations, all these structures are formed according to proposed coloumn geometry.

4. Waves and Wave Length in First and Second Dimension

To support all $\sqrt{S}$ type of photons of EM spectrum, at $\sqrt{S}$ quantum levels in deformation of first dimension of **length,** EM waves are formed in space matrix, while at $\sqrt{S}$ quantum levels, it formes matter waves for all elementary particles in deformation of two dimensions of **length and breadth of unit space matrix.**

Photons and all elementary packets are **not point masses, but in fact they are waves** as they all spread and distribute their mass energy on all its apex bound treos in wave length, to produce EM waves and Matter waves respectively, on their gradually decreasing wave lengths.

This length of spread of packet, in first dimension on its **wave length reduces from S bound treo** (wave length of unit photon) **to $\sqrt{S}$ bound treos** (to wave length of gamma Photon), while in second dimension the spread of elementary particle is on respective RC wave length, which further reduces from **$\sqrt{S}$ bound treos** of one-unit Electron, **to just 1 bound treo** of one-unit mass.

With increasing mass-energy at each next quantum level, the **contraction of mass energy packets** on its reducing wave length continues both in first and second dimension and finally S^2 free treos or S quanta mass-energy of unit mass, contracts and exert its total load **on one graviton at its gravitational centre, which is supported by S^2 kinetons in one graviton coloumn.**

At any quantum level, with reducing wavelength (and increasing mass-energy in packet) the exerted load at each apex bound treo in wave length increases in free treo square (n^2 **free treos**) in **first dimension** and with reducing RC wave length in **second dimension** at each apex bound treo the load exerted is quanta square (n^2 **quanta**), at n^{th} quantum level.

Number of layers in one kinetic coloumn × number of kinetic coloumns one on each apex bound treo forming one EM wave = S number of layers in one EM wave in first dimentional deformation.

Thus total number of quanta mass energy in any photon packet is supported by equal number EM waves, which form in **one second (Action of S layers in one EM wave × number of quanta in packet). S kineton energy is ground energy of one dimensional deformation.**

The mass-energy of any elementary particle packet in free treos × RC wavelength of its matter wave in terms bound treos is equal to S^2. (See page 99) S^2 **kinetic energy is ground energy of two dimensional deformation.**

In one orbitum, by its S vibrations **one layer rotating in S planes** support one quanta mass energy. Total number of orbitums are always equal to the number of quanta mass energy in packet and thus to support packet for **one second (Action of S layers in one orbitum× number of quanta in packet).**

This indicates, that this S^2 kinetic energy is also present on one kinetic coloumn layer at each quantum level of this one graviton coloumn, where one matter wave utilizes this S^2 kinetic energy.

In first dimension all of the rotating sub kinetic coloumns **jointly** produce one transverse EM waves, @ **one EM wave per quanta mass energy per second**, while in second dimension **each orbitum rotating @ once per second (one layer rotating S times in S planes by S vibrations), can individually support one quantum mass energy.**

(a) Common formula, to calculate length of spread in first and RC wave length in second dimension. One sheet of unit Space matrix is woven by S rows of S bound treos which are placed both horizontally and vertically.

With increase of each quantum of mass energy, the photon packet contracts on **decreasing length of deformation of horizontal rows**, along with their increasing slope in 1st dimension, and in second dimension with gradually **increasing depth of pit and reduction of its circumference** (as each circular layer forming this pit gradually contracts on 2 RC wave length x π).

In both dimensions the **LENGTH OF SPREAD OF MASS ENERGY PACKET ON SPACE MATRIX** is decided according to one common formula for first and second dimension.

Length of spread of packet on number of (apex) bound treos = S bound treos length/number of quanta mass energy in this Packet.

(And this formula is valid in both dimensions; from one–unit photon mass to one–unit mass.)

One **photon with spin 1 will spread on its one EM wave and will perform one rotation to regain its orientation.**

At any quantum level on which mass energy of packet spreads and distributes equally in second dimension is its RC wave length (1/2 of wave) on S number of bound treos/number of quanta mass energy in electron or in elementary particle.

In deformation of second dimension if 'n' is Reduced Compton wave length (where 'n' is half of the length of its wave), while this wave is formed on 2n–1 kinetons; and (2n–1) × π is angular momentum of this mass energy packet which decides the circumference of orbit and Compton wave length of wave.

One Electron (or all packets in second dimension) will spread on its half wave along its RC wave length and with ½ spin, it will perform 2 rotation to regain its orientation on 2 RC wave length while revolving and forming its one matter wave.

(b) We will verify all statements with the example of Wave length of unit electron

i. *Wavelength of (0.51 MeV) Electron*

= *0.38615926796 × 10^{-12} m (conventional value)*

=23.797258 × 10^{21} Bound treos length (after conversion).

This can also be calculated by this new formula

S bound treo distance/√S quanta mass energy in electron packet

1.85485844 × 10^{43} bound treo distance/0.779450536 × 10^{21} quanta electron (of √S quanta) = **23.797258 × 10^{21} bound treos length**

ii. *Compton wavelength of (0.51 MeV) electron*

2.42728683 (11) × 10^{-12} m Codata 2018 [Ref.51]

=**1.50179695 × 10^{23} bound treo distance (after conversion).**

(2 × Reduced Compton wavelength, when multiplied by π, it calculates circumference of circle or its Compton wavelength)

2 × 23.797258 × 10^{21} bound treos length× π = 1.50179695 × 10^{23} bound treo length

iii. **Reduced Compton wavelength × mass-energy of electron in free treos = S^2**

23.797258 × 10^{21} Bound treos wave length × 1.44000857 × 10^{64} free treos in unit electron packet= 3.4 × 10^{86} = S^2 **Ground energy of second dimension.**

Inferences

a. If Reduced Compton wavelength is **n**

b. 2 × Reduced Compton wavelength = **2n-1 bound treos on which it forms one wave; in a arc in 1/3ʳᵈ circumference of circle.**

c. While 2n-1 bound treos × π = **Compton wavelength is circumference of this circle.**

5. Frequency in First and Second Dimension

The **number of quanta** as mass energy in any packet, decide the **FREQUENCY of wave both in first and second dimention,** which is also equal to the **number of bound treo layers** in each kinetic coloumn (or in each shell in second dimension), which are formed at each apex bound treo along the reduced Compton wave length of packet.

The orbital speed of revolution of any body (electron, elementary particle, planet or sattelite) is always equal to the frequency of its matter wave produced at this particular quantum level in its orbit.

6. Deformation in One Dimensions of Length

(To support All types of Photon packets from unit Photon to gamma photon.)

We know that the field of EM force is formed by transverse EM waves in which electrical and magnetic vectors are placed at 90 degrees to each other. According to proposed model the **electrical vector is manifested by free treos in packet, while vertically placed revolving and supporting kinetic coloumns generate magnetic field.**

The load of any photon packet is shared on space matrix. Any photon packet will spread along its **wave length** (of S bound treo/number of quanta in photon packet) and the load of any number of free treos (according to its photon packet density) at each apex bound treo, is supported by one sub kinetic coloumn having equal number of kinetons.

At **first quantum level (n=1),** one quantum mass energy (S free treos in unit photon) spreads on S number of apex bound treos along its wave length. Now at each apex bound the **load is one free treo,** which is supported by one kinetic coloumn having one kineton in its first layer.

At **second quantum level (n=2),** two quanta mass energy photon packet (double the mass energy) spreads on S/2 apex bound treos in wave length (spreads on half length) and exert a 'load' of 4 or 2^2 free treos. At each apex bound in wave length, the load of 4 free treo is supported by 4 kinetons in one kinetic coloumn which have one kineton in its first layer and three kinetons i.e. $2 \times 2 - 1 = 3$ **(2 n – 1)** in its second layer.

At **third quantum level (n=3),** three quanta photon packet spreads on S/3 apex bound treos in wave length and exert a 'load' of 9 or 3^2 free treos at each apex bound treo, which is supported by one kinetic coloumn of 9 kinetons, having one kineton in its first layer, three kinetons in its second layer and five kinetons i.e. $3 \times 2 - 1 = 5$ **(2 n – 1)** in its third layer. Thus, total 3^2 kinetons $(1 + 3 + 5 = 9)$ are in this three layered kinetic coloumn.

Similarly, at **Fourth quantum level (n=4),** for Four quanta photon packet spreads on S/4 apex bound treos in wave length and exert a 'load' of 16 or 4^2 free treos at each apex bound treo, which is supported by one kinetic coloumn of 16 kinetons, having one kineton in its first layer, three kinetons in its second layer, five kinetons in third and seven kinetons i.e. $4 \times 2 - 1 = 7$ **(2 n – 1)** in its fourth layer. Thus, total 4^2 kinetons $(1+3+5+7 = 16)$ are in this four layered kinetic coloumn.

Similarly, at all **√S quantum levels (n=√S),** for √S quanta photon packet one √S layered kinetic coloumns is formed, at each apex bound treo along its √S bound treos wave length.

To support the increasing load of n^2 free treos at n^{th} quantum level, one new layer of **2n –1 kinetons is added** in each Sub kinetic coloumn (present on each apex bound treos in wave length) to complete one n layered sub kinetic coloumn, having equal number of n^2 kinetons (square of the number of layers in coloumn).

1.   **First Quantum Level of First Dimension (Deformation of 1 Row of S Bound Treos)**

S free treos in unit photon packet spreads on S bound treos length on one layer of unit space matrix. It will exert a uniform load of one free treo to be supported by S number of one layered sub kinetic coloumns in this wave length. Thus, total of S free treos in unit photon packet are supported by **one row of S kinetons** of unit space matrix (Fig 16).

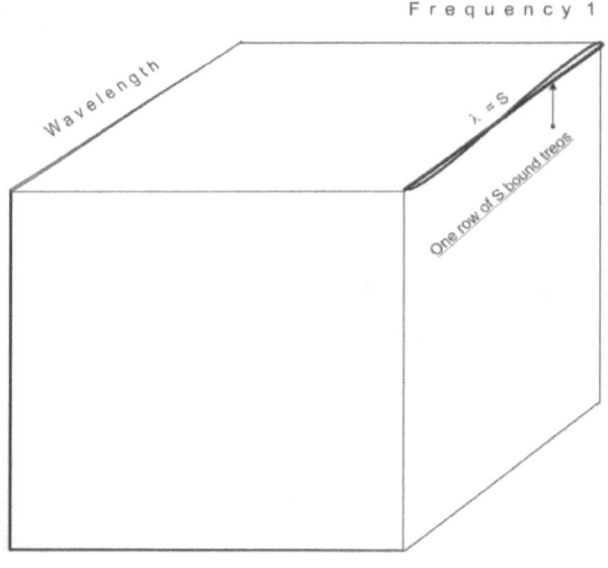

First Quantum Level of first dimention

(Diagrammatic Representation)

Figure 16: First quantum level in deformation of first dimension in length

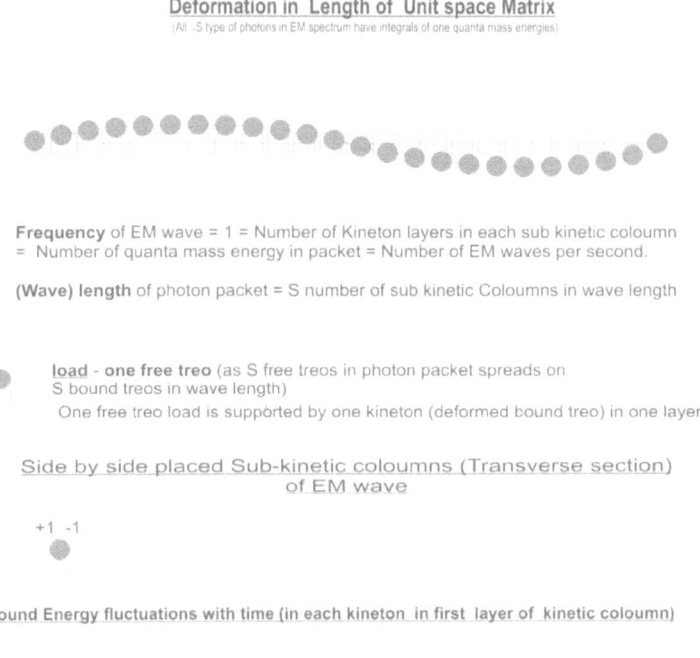

One quanta (S free treos) 'Unit photon packet'
Deformation in Length of Unit space Matrix
(All S type of photons in EM spectrum have integrals of one quanta mass energies)

Frequency of EM wave = 1 = Number of Kineton layers in each sub kinetic coloumn = Number of quanta mass energy in packet = Number of EM waves per second.

(Wave) length of photon packet = S number of sub kinetic Coloumns in wave length

load - one free treo (as S free treos in photon packet spreads on S bound treos in wave length)
One free treo load is supported by one kineton (deformed bound treo) in one layer

Side by side placed Sub-kinetic coloumns (Transverse section) of EM wave

+1 -1

Ground Energy fluctuations with time (in each kineton in first layer of kinetic coloumn)

diagramatic representation

Figure 17: First quantum level in deformation of first dimension in length

One EM wave is shown in Fig 17. One free treo is supported by one kineton in one layered sub kinetic coloumns in wavelength.

2. Second Quantum Level of First Dimension (addition of 3 Rows in Deformation)

Two quanta (or 2 × S free treos) is mass energy of two quanta photon packet. And according to formula S bound treo/2 quanta mass energy in packet= **S/2 bound treos is wave length.**

Thus, 2 × S free treos in photon packet, uniformly spread on S/2 bound treo wave length will exert a uniform load of 4 free treo at each of S/2 apex bound treos in wave length, where its 4 free treo load is supported by 4 Kinetons (1 + 3 kinetons) or 2^2 kinetons, in each of two layered S/2 sub kinetic coloumns (Fig. 19), Thus, to form all sub kinetic coloumns in wave length, 1 row of S bound treos is already deformed and now **3 new rows of S/2 bound treos will deform**. (Fig. 18)

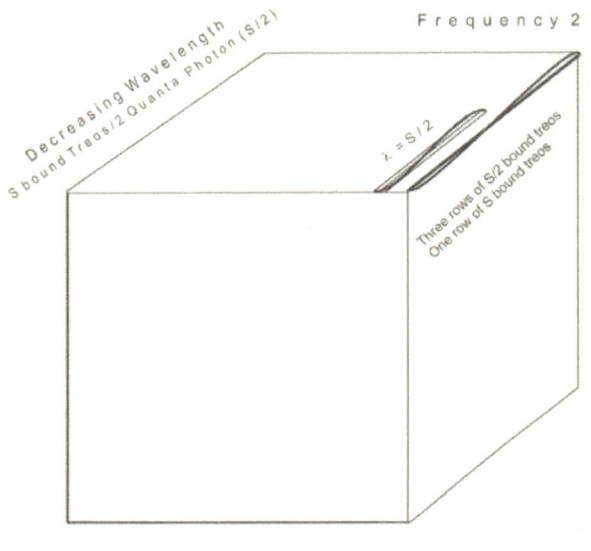

Second quantum level of First Dimension

(Diagrammatic Representation)

Figure 18: Second quantum level in deformation of first dimension in length: additional deformation of 3 rows of S/2 bound treos is required to form all sub kinetic coloumns.

While **2 quanta mass energy** is in this two quanta photon packet and **2 layers are in its each sub kinetic coloumn, 2 is frequency** of this wave and with 2 unit angular momentum it will form 2 EM waves in one second.

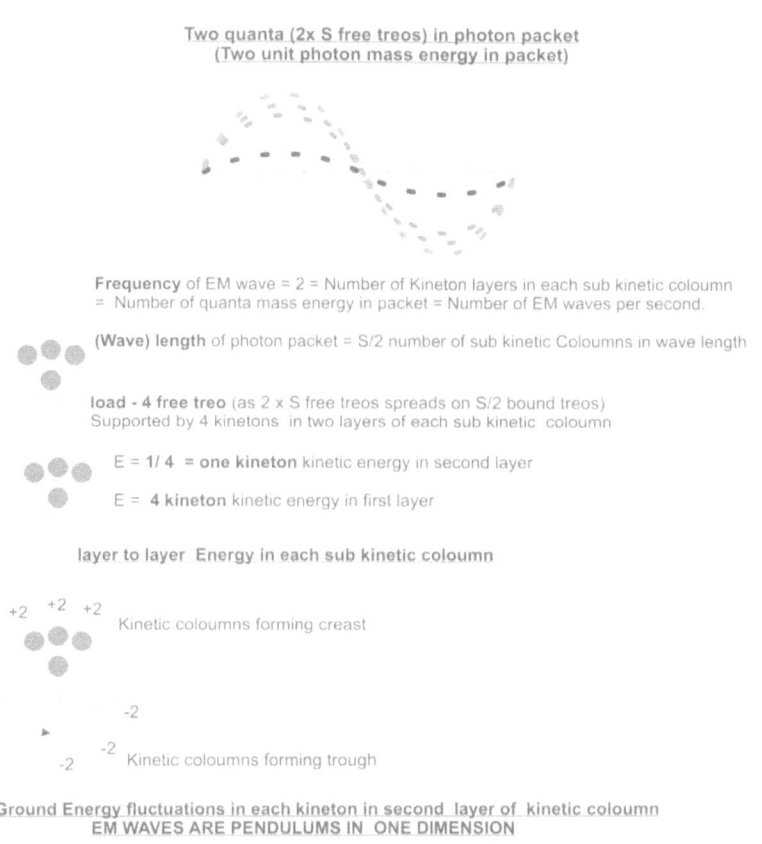

Figure 19: Second quantum level in deformation of first dimension in length

One wave is shown in Fig. 19. 4 free treos are supported at each apex bound treo by 4 kinetons in two layers in each sub kinetic coloumn.

3. Third Quantum Level of First Dimension (Addition of 5 Rows in Deformation)

Three quanta (3 × S free treos) is mass energy of three quanta photon packet. And according to formula S bound treo/3 quanta mass energy in packet = S/3 bound treos is wave length.

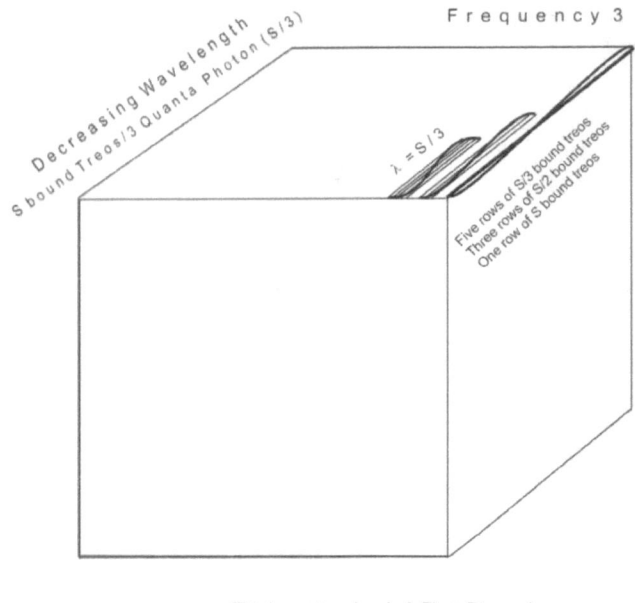

Third quantum level of First Dimension

(Diagrammatic Representation)

Figure 20: Third quantum level in deformation of first dimension in length: additional deformation of 5 rows of S/3 bound treos is required to form all sub kinetic coloumns.

One row of S bound treos and 3 rows of S/2 bound treos are already deformed; Now 5 new rows of S/3 bound treos are added and thus at each apex bound treo along its S/3 bound treo wave length, it forms one sub kinetic coloumn of 3^2 kinetons (1 + 3 + 5 = 9 kinetons), in its 3 layers. (Fig. 20)

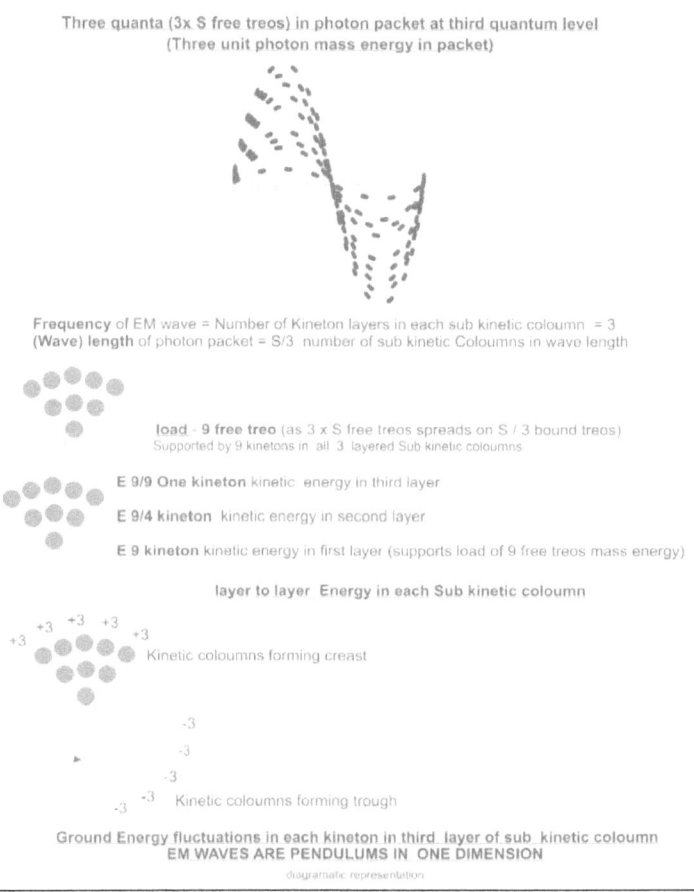

Three quanta (3x S free treos) in photon packet at third quantum level
(Three unit photon mass energy in packet)

Frequency of EM wave = Number of Kineton layers in each sub kinetic coloumn = 3
(Wave) length of photon packet = S/3 number of sub kinetic Coloumns in wave length

load - 9 free treo (as 3 x S free treos spreads on S / 3 bound treos)
Supported by 9 kinetons in all 3 layered Sub kinetic coloumns

E 9/9 One kineton kinetic energy in third layer

E 9/4 kineton kinetic energy in second layer

E 9 kineton kinetic energy in first layer (supports load of 9 free treos mass energy)

layer to layer Energy in each Sub kinetic coloumn

+3 +3 +3
+3   +3
Kinetic coloumns forming creast

-3
-3
-3
-3 -3   Kinetic coloumns forming trough

Ground Energy fluctuations in each kineton in third layer of sub kinetic coloumn
EM WAVES ARE PENDULUMS IN ONE DIMENSION
diagramatic representation

Figure 21: Third quantum level in deformation of first dimension in length

Formation of one EM wave at third quantum level is shown in Fig. 20. All sub kinetic coloumns are formed, by 1 row of S bound treos, 3 rows of S/2 bound treos which are already deformed and **now additional 5 rows of S/3 bound treos will deform.**

Thus total 3 × S free treos in three quanta photon packet, spread uniformly on S/3 bound treo wave length and exert a uniform load of 9 free treo at each of these S/3 apex bound treos, where its 9 free treo are supported by 9 Kinetons in each sub kinetic coloumn (9 kinetons = 1 + 3 + 5 kinetons in three layers). (Fig.21)

While 3 quanta mass energy is in this three quanta photon packet and **3 layers are in its each sub kinetic coloumn,** 3 is frequency of this wave and with 3 unit angular momentum, it will form 3 EM waves in one second.

4. Fourth Quantum Level of First Dimension (Addition of 7 Rows in Deformation)

Four quanta (or 4 × S free treos) is mass energy of four quanta photon packet. And according to formula S bound treo/4 quanta mass energy in packet and thus S/4 bound treo is wave length.

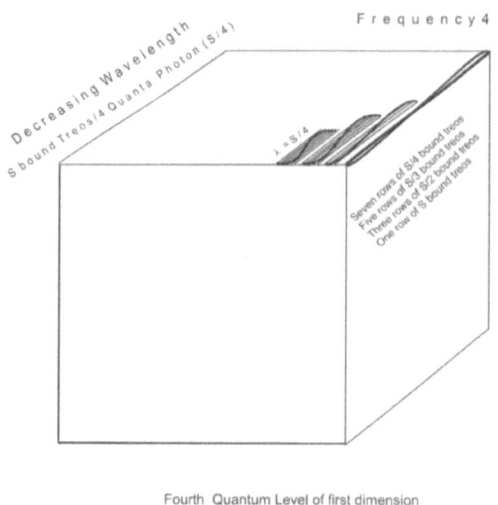

Fourth Quantum Level of first dimension
(Diagrammatic Representation)

Figure 22: Fourth quantum level in deformation of first dimension in length: additional deformation of 7 rows of S/4 bound treos is required to form all sub kinetic coloumns.

One row of S bound treos, 3 rows of S/2 bound treos and 5 rows of S/3 bound treos are already deformed/contracted. **Now 7 rows of S/4 bound treo length are added** and thus at each apex bound treo along S/4 wave length one sub kinetic coloumn form of 1+3+5+7= 16 kinetons in 4 layers. (Fig. 22)

Thus total 4 × S free treos in photon packet, uniformly spread on S/4 bound treo wave length, exert a uniform load of 16 free treo at each of these S/4 apex bound treos in wave length where its 16 free treo are supported by 16 kinetons in each sub kinetic coloumn (16 kinetons = 1+3+5+7) or 4^2 kinetons, in four layers. (Fig.23)

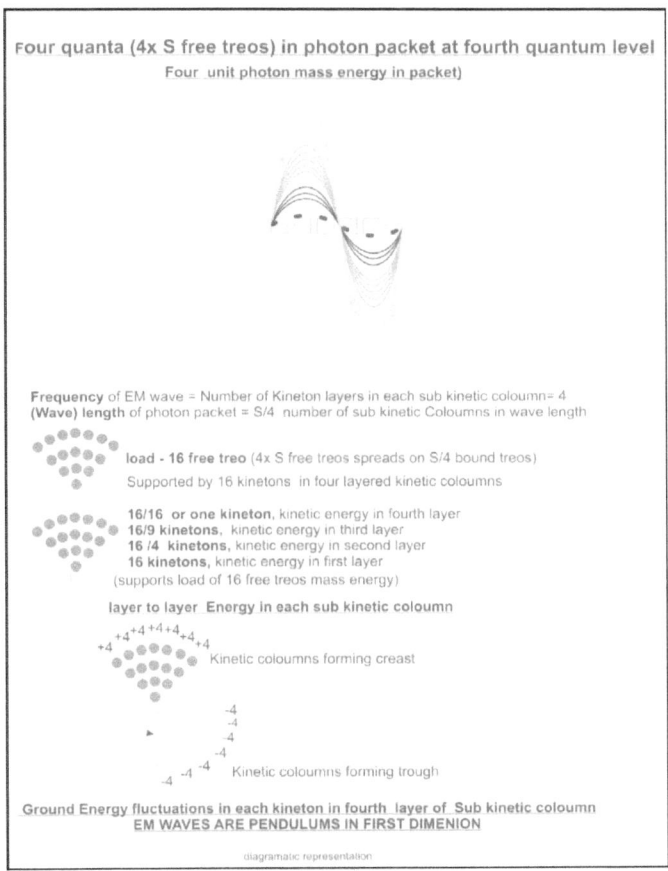

Figure 23: Fourth quantum level in deformation of first dimension in length

One EM wave, with one kineton in first layer, 3 kinetons in second, 5 kinetons in third layer and 7 kinetons are in each sub kinetic coloumn, which are present at each apex bound treo, in S/4 bound treo wave length; (Fig. 23).

While **4 quanta mass energy** is in this four quanta photon packet and **4 layers are in its each sub kinetic coloumn**, 4 is frequency of this wave and with 4 unit angular momentum, it will form 4 EM waves in one second.

5. **Last or √S Quantum Level of First Dimension (Addition of New 2√S-1 Rows)**

The gamma photon packet at last quantum level is of √S quanta (or √S×S free treos) mass energy. And according to formula S bound treo/√S quanta mass energy in packet; S/√S = **√S bound treos wave length.**

If we see total deformation by √S quanta (or √S × S free treos) mass energy, one row of S kinetons, 3 row of S/2 apex bound treos wave length, 5 rows of S/3 bound treos and 7 rows in S/4 bound treo length and to continue, so on ..... 2n−1 row of S/n bound treos are added at each nth quantum level and get deformed, finally **new 2 √S-1 rows, each of √S = (S/√S) bound treos** length are added at last √S quantum level. (Fig.24)

√S × S free treos in gamma photon packet uniformly spreads on √S bound treo wave length and thus exert a uniform load of S free treo or one quantum at each of √S apex bound treos in wave length. Where these S free treo are supported by S Kinetons in each sub kinetic coloumn. Each sub kinetic coloumn is made up of **S kinetons, in √S layers.**

While **√S quanta is mass energy** in biggest photon packet and **√S layers are in its each sub kinetic coloumn, √S is frequency** of this wave and thus with √S units angular momentum, √S waves will form in one second.

Deformation of first dimension in one unit time (of 1 second).

At √S[th] quantum level for gamma photon it is the deformation of all **S rows** of unit space matrix, jointly at all quantum levels **in deformation of first dimension in length.**

As per coloumn geometry 2n-1 rows are deformed at each nth quantum level; therefore in complete deformation of first dimension, from first to $\sqrt{S}$ quantum levels, total S rows (n^2) will be deformed.

As observed all Photons ranging from 1 to $\sqrt{S}$ quanta mass energy forms equal number of 1 to $\sqrt{S}$ waves in one second, to support itself for one second.

The deformation of **S kinetons in 1 row** for one quanta photon packet, will form one wave in one second at first quantum level. One quanta photon packet takes 1 rotation (and form 1 wave) in one second to be supported for one second.

At second quantum level (n=2) **three full rows (2n-1) each having S kinetons** will be used to form two waves of S/2 bound treo wave length in one second. Two quanta photon packet will take 2 rotations (and forms 2 waves) in one second to be supported for one second.

At third quantum level (n=3) **five full rows (2n-1) each having S kinetons** will be used to form three waves of S/3 bound treo wave length in one second. Three quanta photon packet will take 3 rotations (and forms 3 waves) in one second to be supported for one second.

At fourth quantum (n=4) level **full seven rows (2n-1) each having S kinetons** will be forming four waves of S/4 bound treos wave length in one second. Four quanta photon packet will take 4 rotations (and forms 4 waves) in one second to be supported for one second.

Thus gamma photon of $\sqrt{S}$ quanta mass energy is supported in one second, by $\sqrt{S}$ rotations of this photon packet on $\sqrt{S}$ waves. e.g. The gamma photon will remain supported on its $S/\sqrt{S}$ wave length **by $S/\sqrt{S}$**

vibrations, for S/√S second time only on its one wave. To support it for S vibrations i.e. for one second, total S vibrations (= S/√S vibrations in one wave × √S waves in one second) for √S waves it will require **full 2√S − 1 rows each having S kinetons.**

√S EM waves of gamma photon, (together with moving deformation at lower quantum levels) will use **total S kinetons in all S rows of first dimension, to support √S quanta mass energy of this gamma photon in unit time of one second.**

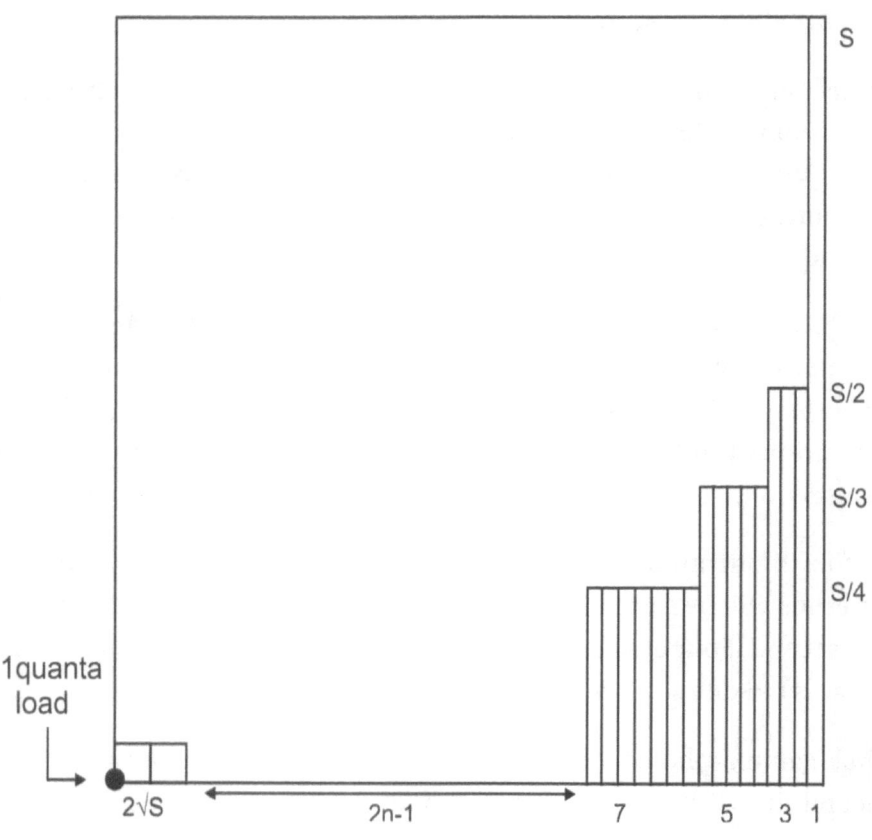

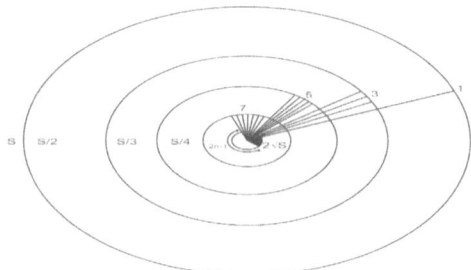

Figure 24: √S quantum level in deformation of first
dimension in length: additional deformation of **2√S-1 rows of √S bound treos**
each are required to form all sub kinetic coloumns (not to scale)

7. Deformation in Two Dimensions of Length and Breadth

Packet of one unit electron is √S quanta mass energy (or √S ×S free treos). At each next quantum level, of second dimensional deformation, this √S quanta mass energy of one unit electron packet increases (**as one unit**) up to √S quantum level to produce all energetic electrons, all elementary particles, nucleons, atoms, molecules and body up to one unit mass.

These mass energy packets of all elementary particle get uniformly spread and exert a equal 'load' on all apex bound treos in its succesively reducing RC wave length (thus gradually contracting). To neutralize this load on each apex bound treo, one shell (kinetic coloumn) at each apex bound area will form as one bound treo thick disc, generated from one layer of unit space matrix.

Thus this RC wave length (S bound treos/n quanta mass energy in packet), gives S/n bound treo height to this vertically placed packet.

In start of deformation of second dimension, the mass energy packets at 1st, 2nd, 3rd and 4th quantum level will gradually shrink alongwith reducing RC wave length of √S, ½√S, ⅓√S, ¼√S apex bound treos length.

In deformation of second dimension, the 'coloumn geometry' will remain same (**2n-1 units in its any nth layer, while n^2 are in n layered coloumn**) but **sub kinetic coloumn, layers of sub kinetic coloumn** and **kineton** of first dimension are replaced by **shell, sub shells** and **orbitum** respectively in second dimensional deformation, as they are formed in **two–dimensional geometry** of second dimension in one plane.

Breadth of particle

At any nth quantum level **2n-1 orbitums** (each of S kinetons) in nth sub shells of each shell are added (in place of **2n-1 kinetons** in any newly added nth layers in sub kinetic coloumn). Thus In each shell (kinetic coloumns), one new sub shell added at each next quantum level, will also **add √S bound treos layers** in its increasing radius, with the addition of one new row of 2n-1 small squares (as each square of √S bound treos on its two sides also have √S bound treos along its diagonal). (see Fig 25).

Thus If total **radius** of disc of shell, **is 'n' number of bound treos layers** at any quantum level; then the **number of quanta mass energy** in this packet is also 'n'; also 'n' is the **Frequency** of wave.

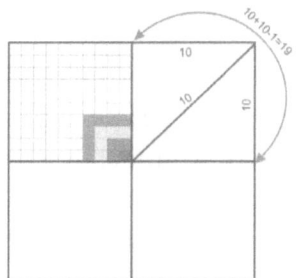

Figure 25: (e.g. for understanding of unit space matrix geometry); A small area of **undeformed space matrix** can be divided in four squares, representing all four directions in two dimensional geometry. Each portion will have 100 blocks (small squares) which will be arranged like above diagram in space matrix. Each have 10 blocks in length, 10 in breadth **and 10 in diagonal**. Two sides in each row deformed diagonally from one corner will have (2n-1 blocks) e.g. at periphery it has total 19 blocks (= 10 +10 − 1).

The **unit space matrix (it is for calculation) can be counted from any point on space matrix where the load is applied.** It will start from the point of exerted load as one upper corner of one unit space matrix. With reducing wave length from √S bound treos of a unit electron to one bound treo of 'one unit mass'.

With one shell formed in one sheet, the wave involves √S sheets of one unit space matrix, on its RC wave length at first quantum level for one unit electron mass (and in gradually reducing number of shells and sheets involved with decreasing wave length at each next quantum level for all elementary particles) up to just one last sheet at last √S quantum level, at one unit gravitational centre for one 'unit mass'. **Thus in total it will involve all S sheets, for formation of √S quantum levels in depth in second dimensional deformation**

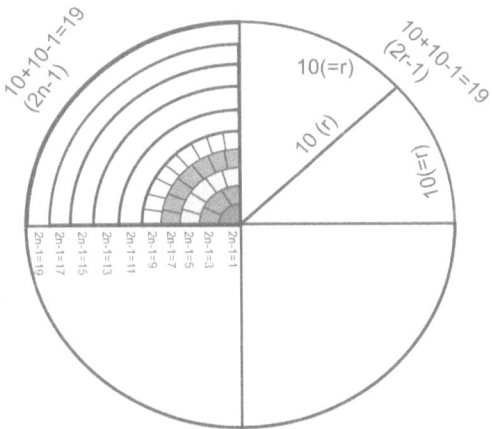

Figure 26: (e.g. for understanding of unit space matrix geometry); This one portion of one sheet of space matrix shown in Fig. 25 contract to form **kinetic coloumn of shell** as shown, and it will have 100 kinetons (which came from 100 small squares) and will be arranged like above diagram. 1+3+5+7+9+11+13+15+17+19 =100 in 10 layered coloumn, (or 2n-1 squares in each row and n2 are in any n layered coloumn).

1. First Quantum Level

'√S quanta mass energy' in unit electron packet exert one quantum load. This 'one quantum load' is exerted at each of √S apex bound treos in its RC wave length, and is supported by each of √S shells, which are vertically placed and form in √S sheets of unit space matrix.

Each rotating shell has one sub shell and one orbitum (present in all shells in wave length and will jointly from **1 s orbit**) made up of S kinetons, which comes from S bound treos in first square deformed (of √S bound treos are in its one side of square and S bound treos in its area) present in first row. Fig 29 page 123.

2. Second Quantum Level

2 × '√S quanta mass energy' in two unit electron packet exert 4 quantum load or $(2)^2$ quanta load, at each of √S/2 apex bound treos in its RC wave length according to its 'electron packet density' (as mass energy doubles on half wave length).

This 4 quanta load at each of √S/2 apex bound treos in its RC wave length is supported by √S/2 shells which are placed vertically and are formed in √S/2 sheets of one unit space matrix, one new sub shell will be added.

Each rotating shell has two sub shells; with one orbitum in first sub shell and 3 orbitums in second sub shell (while each orbitum is made up of S kinetons), jointly form 3 p orbits.

First sub shell with one orbitum is at √S bound treo radius, while second sub shell has all its 3 orbitums at 2 × √S bound treo radius. One orbitum of first sub shell gets its S kinetons from one square (of √S bound treos are in its one side of square and S bound treos in its area) in first row, while all 3 orbitums of second sub shell gets 3 S kinetons from 3 squares present in second row deformed. (Fig. 33, page 127)

3. Third Quantum Level

3 × '√S quanta mass energy' in three unit electron packet exert 9 quantum or $(3)^2$ quanta load, equal to its 'electron packet density' (as mass energy triples on one third wave length). This 9 quanta load at each of √S/3 apex bound treos in its wave length is supported by each of √S/3 shells (in RC wave length) which are placed vertically and are formed in √S/3 sheets of unit space matrix.

Each rotating shell has three sub shells; with one orbitum in first sub shell, 3 orbitums in second sub shell and 5 orbitums in newly added third sub shell will jointly form 5 orbits.

First sub shell with one orbitum is at √S bound treo radius, while second sub shell has all its 3 orbitums at 2 × √S bound treo radius and third sub shell has all its 5 orbitums (present in all shells in wave length will jointly from **5d orbits**) at 3 × √S bound treo radius. One orbitum of first sub shell gets its S kinetons from one square in first row, while all 3 orbitums of second sub shell gets 3 S kinetons from 3 squares present in second row deformed and all 5 orbitums of third sub shell gets 5 S kinetons from 5 deformed squares present in third row (Fig 37, page 131).

4. Fourth Quantum Level

4 × '√S quanta mass energy' in four unit electron packet exert 16 quantum or $(4)^2$ quanta load, equal to its 'electron packet density'. This 16 quanta load at each of √S/4 apex bound treos in its wave length is supported by each of √S/4 shells (in RC wave length) which are placed vertically and are formed in √S/4 sheets of unit space matrix.

Each rotating shell has four sub shells; with one orbitum in first sub shell, 3 orbitums in second sub shell, 5 orbitums in third sub shell and 7 f orbitums are in newly added fourth sub shell.

With **every increase of '√S quanta mass energy' (in one unit electron) at each next quantum level, one new sub shell of √S bound treo layers are added.** Thus concentric √S bound treo layers increases at each next n^{th} quantum level, and with one new sub shell added, while the total radius of shell will become n x √S bound treo layers.

From 1 small square of √S bound treos sides, present at corner of one sheet of unit space matrix S bound treos from its S bound treo area, generates 1 S kinetons which form one orbitum. (Fig. 30, page 124)

At second quantum level with the deformation of new √S bound treos layers (of second row) which increases diagonally it will form second sub shell, and now 4 S bound treos in 4 S bound treo area (1 square + 3 squares) in each of shell get converted in to 4 S kinetons to form 4 orbitums. (Fig. 34, page 128)

At third quantum level with the deformation of new √S bound treos layers diagonally (of third row) it will form third sub shell, and now 9 S bound treos in 9 S bound treo area (1 square + 3 squares + 5 squares) in each of shell get converted in to 9 S kinetons to form 9 orbitums. (Fig 38, page 132)

At Fourth quantum level with the deformation of new √S bound treos layers diagonally (of fourth row) it will form fourth sub shell, and now 16 S kinetons from 16 S bound treo area (1 square + 3 squares + 5 squares + 7 squares) in each shell get converted in to 16 S kinetons to form 16 orbitums (Fig 42, page 136).

Thus the area of each shell increases at next quantum level by addition of (2n-1 quanta) i.e. 3S bound treos area from 3 squares in second sub shell, 5S bound treos area from 5 squares in third sub shell and 7S bound treos area from 7 squares in fourth sub shell from which it forms additional 3, 5 and 7 supporting orbitums in 2nd, 3rd and 4th sub–shells in each shell.

All identical energy orbitums 1+3+5+7 orbitums present in s, *p, d, f* sub shells of each shell (one below other in all shells) at fourth quantum level will jointly form 16 orbits, at this fourth quantum level. (Fig 42)

We can note that at 1st, 2nd, 3rd and 4th quantum level, **one quantum**, 2^2 **Quanta**, 3^2 **Quanta**, 4^2 **Quanta** load are respectively present at each apex bound treo in RC wave length of packet.

At first quantum level **1 quantum load** is supported by **1** orbitum in one first sub shell, at second quantum level, with 3 new orbitums added in one new second sub shell, $1 + 3$ = total 4 orbitums supports 4 quanta or 2^2 **Quanta load**, at third quantum level it adds 5 orbitums in one new third sub shell, $1 + 3 + \mathbf{5}$ = total 9 orbitums in three sub shells support 9 quanta load or 3^2 **Quanta load** and then seven orbitums in fourth new sub shell $1 + 3 + 5 + \mathbf{7}$ = total 16 orbitums are in first, second, third and fourth sub shells of each shell (which are placed at each apex bound treo along its wave length) and supports 16 quanta load or 4^2 **Quanta load.**

One orbitum will support one quantum mass energy by its one rotation in one second.

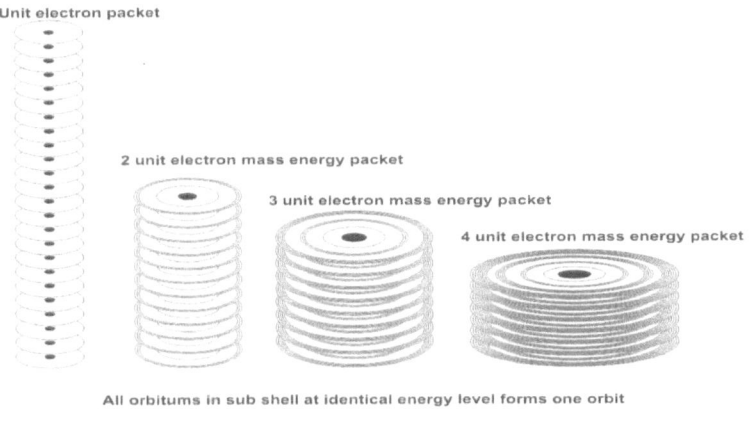

Unit electron packet

2 unit electron mass energy packet

3 unit electron mass energy packet

4 unit electron mass energy packet

All orbitums in sub shell at identical energy level forms one orbit

Formation of $1\,s + 3\,p + 5\,d + 7\,f$ orbits with increasing deformation at 1st, 2nd,3rd and 4th quantum levels of second dimension

Figure 27: *1s, 3p, 5d* and *7f* orbits are formed in four sub shells in each shell per Coloumn geometry at first, second, third and fourth quantum level in deformation of Second dimension.

At subsequent each next n^{th} quantum level the elementary particle packet which condenses, will have n √S quanta mass energy (thus all elementary particle described and not described in standard model, are formed), and n^2 quanta is load (with the increasing mass and decreasing RC wave length) at each apex bound treo which is supported by addition of 2n−1 orbitums in n^{th} new sub shell, in each shell which are present on each apex bound treo along its S/n bound treos RC wave length.

At n^{th} quantum level with the deformation of new √S bound treos layers diagonally it will increase the area of shell by 2n-1 × S bound treos, which will add 2n-1 × S kinetons for 2n-1 new orbitums, which are formed in one new n^{th} sub shell.

One unit mass (Planck mass; 2.176 × 10^{-8} Kg) is maximum load[17] (roughly the weight of flea egg) which can be supported at one bound treo on space matrix anywhere in universe at its one unit gravitational centre or **at one graviton by its one graviton coloumn (full kinetic coloumn of second dimension).**

As the successively decreasing RC wave length finally reduces to one bound treo (1 bound treo wave length = S bound treos/S quanta mass energy in one unit mass); for one−unit mass which is supported at one graviton (i.e. one apex bound treo) by one graviton coloumn.

In this graviton coloumn in one last sheeth deformed (1 bound treo RC wave length) at last √S quantum level, it will add one new sub shell with 2√S − 1 (2n-1) orbitums. At this last √S quantum level, in its all √S sub shells (in last S^{th} sheet) toal S orbitums will be in last shell, which are formed by S^2 kinetons from last full sheet of unit space matrix, and

---

17 (A body made up of multiple unit masses exerts a load in square of its unit masses at its gravitational centre, which is supported by equal number of gravitons in one 'electron black hole' in third and in 'one gravitational sphere' in fourth dimension, which forms around its gravitational centre.)

will support S quanta load of unit mass (of S quanta mass energy or of S^2 free treos).

The wave at graviton is of S frequency; so all S orbitums in this last sheet will rotate around graviton, at the speed of light to complete their one rotation in one second. Thus S quanta load is supported by a graviton.

Along with, in deformation at lower quantum levels total $\sqrt{S}$ number of matter waves is present, revolving at gradually reducing speeds, one at each of $\sqrt{S}$ quantum levels with complete deformation of two dimensions.

Theme of Deformation – One free treo is supported by one kineton and thus each quanta mass energy (S free treos) is supported by one orbitum made up of S kinetons. **Each orbitum rotates once in one second to support its load of one quanta mass energy.**

1. First Quantum Level in Deformation of Second Dimension

(Deformation of $\sqrt{S}$ bound treo layers and S kinetons in 1 small square)

The unit Electron spreads its $\sqrt{S}$ quanta mass energy (or $\sqrt{S} \times S$ free treos packet) on $\sqrt{S}$ apex bound treo in its **Reduced Compton wave length (S bound treo/$\sqrt{S}$ quantum mass energy in packet)**. The load is one quantum at each of $\sqrt{S}$ apex bound treo, as it evenly distributes its mass energy in vertically placed packet along its $\sqrt{S}$ RC wave length.

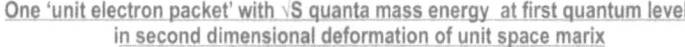

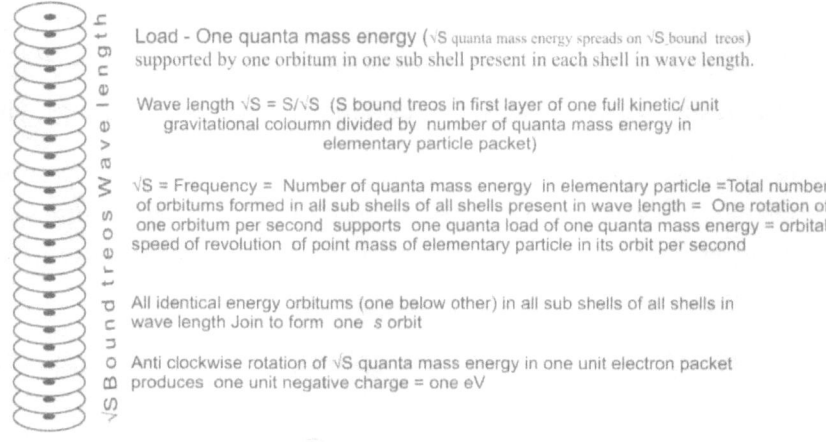

Load - One quanta mass energy (√S quanta mass energy spreads on √S bound treos) supported by one orbitum in one sub shell present in each shell in wave length.

Wave length √S = S/√S (S bound treos in first layer of one full kinetic/ unit gravitational coloumn divided by number of quanta mass energy in elementary particle packet)

√S = Frequency = Number of quanta mass energy in elementary particle =Total number of orbitums formed in all sub shells of all shells present in wave length = One rotation of one orbitum per second supports one quanta load of one quanta mass energy = orbital speed of revolution of point mass of elementary particle in its orbit per second

All identical energy orbitums (one below other) in all sub shells of all shells in wave length Join to form one *s* orbit

Anti clockwise rotation of √S quanta mass energy in one unit electron packet produces one unit negative charge = one eV

Section of matter wave or Structure of one shell with 1 sub shell
& 1 orbitum which support one quanta mass energy

diagramatic representation

Figure 28: Coloumn geometry in Second dimension at first quantum level

√S quanta, **mass energy** in one unit electron deforms √S **bound treo layers** in each shell (kinetic coloumn) at first quantum level, and thus √S **is frequency** of this wave. (Fig. 29)

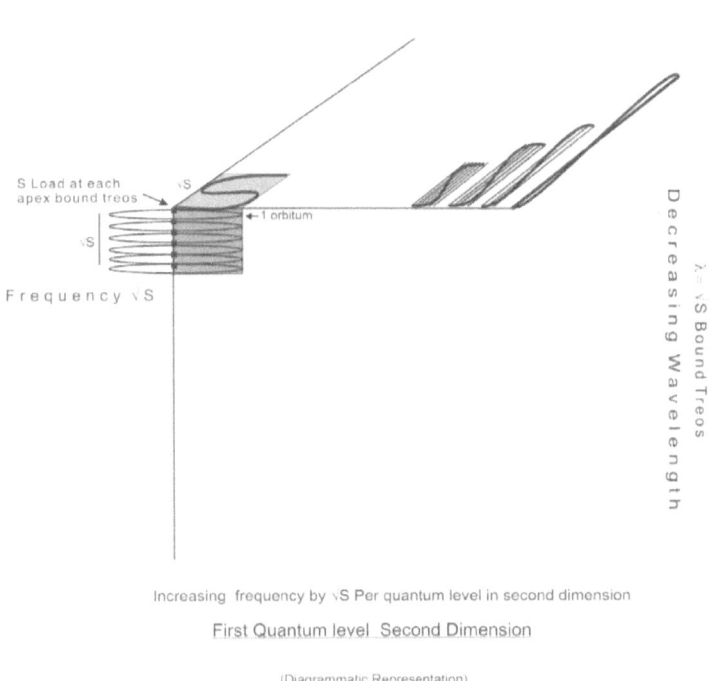

Frequency=Number of Quantum in Packet
Wave length =S bound Treos / √S Quantum in Packet =√S Bound Treos

Figure 29: Coloumn geometry in Second dimension at first quantum level **(side view):** one orbitum forms from S kinetons in a square of √S bound treos and for √S orbitums in wave length, √S such squares will deforms from one corner cube of √S bound treos.

This, S free treos (i.e. one quantum) load is supported by S kinetons of one orbitum which is present at each apex bound treo (in one sub shell of shell) along RC wave length of one–unit electron.

One deformed square has √S bound treos in its one side and also in its diagonal and has area of S bound treos (√S × √S bound treos), which get converted into **S kinetons to form one orbitum.**

This one square is from one corner of one sheeth while √S squares of √S sheets of unit space matrix one below other deforms to form √S orbitums.

One orbitum which form at each of √S apex bound treos in vertical wave length, rotates once in one second (in its sub shell of one shell), and all orbitums one below other form one orbit named as **s orbit**. (Fig. 30)

All these squares forming all orbitums are from √S sheets, thus total deformation is of **one corner cube having all its side of √S bound treos (Fig. 29).**

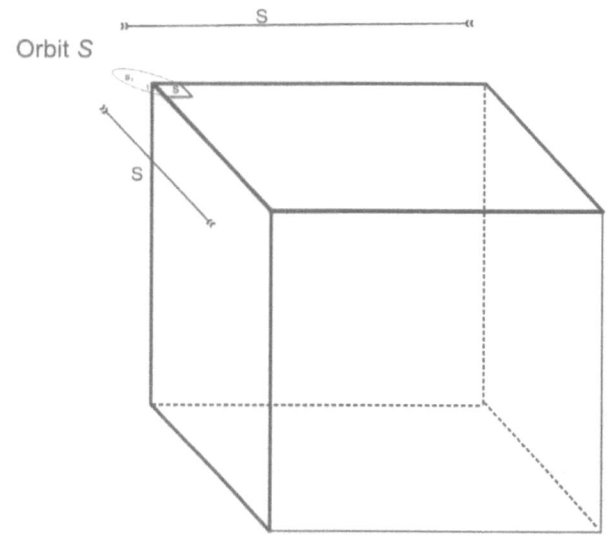

Figure 30: Coloumn geometry in Second dimension at first quantum level

2. Second Quantum level in deformation of second dimension

(addition of new √S bound treo layers diagonally, along with 3 new small squares of second row)

The two–unit Electron packet is of 2 × √S quanta mass energy or 2 (√S × S) free treos. And according to formula √S/2 bound treo is its **RC wave length**. (Fig. 31)

2 'unit electron packet' with 2x√S quanta mass energy at second quantum level in second dimensional deformation of unit space marix

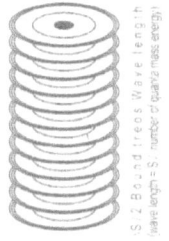

Load - 4 quanta mass energy (as 2x√S quanta mass energy spreads on √S/2 bound treos) supported by (1+3 =) 4 orbitums in each shell

Wave length ≈ S/ 2√S (S bound treos in first layer of one full kinetic / unit gravitational coloumn divided by number of quanta mass energy in energetic electron packet)

2 √S ≈ Frequency = Number of quanta in elementary particle =Total number of orbitums formed in all sub shells of all shells present in wave length = One rotation of each one orbitum per second supports one quanta load of one quanta mass energy per second = orbital speed of revolution of point mass of elementary particle in its orbit per second

All identical energy orbitums (one below other) in all subshells of all shells in wave length Join to form *1s* and *3p* orbits

Anti clockwise rotation of 2 √S quanta mass energy in 'two unit electron packet' produces two unit negative charge = two eV

Section of matter wave or Structure of one shell with 2 sub shells & (1+3) 4 orbitums

diagramatic representation

Figure 31: Coloumn geometry in Second dimension at second quantum level

Thus two–unit electron mass energy packet exert load of 4 quanta i.e. 2^2 **quanta** (as double mass energy spread in half wave length) at each apex bound treo, which is supported by 1+3 orbitums in two sub shells of each shell.

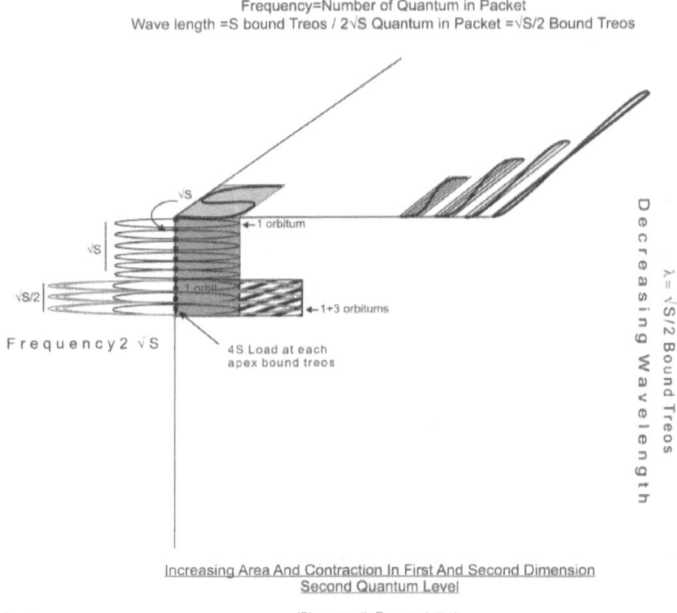

Figure 32: Coloumn geometry in Second dimension at second quantum level (**side view**) : one orbitum forms from S kinetons in a square of √S bound treos and for additional 3 orbitums in second sub shell, 3 such squares from second row deforms. For all orbitums in wave length such 1+3 squares deform are from two rows, in all √S/3 sheets which deform below first quantum level.

1 square in first row (of √S bound treo layers) forms S kinetons and 3 squares in second row (of √S bound treos layers, which increases diagonally) of kinetic coloumn adds 3S kinetons from one sheet, thus form total 1+3 orbitums in two sub shells (from two row) of each shell at each of √S/2 apex bound treo in wave length and thus vertically placed √S/2 more sheets of one unit space matrix are involved just below the deformation of first quantum level.(Fig. 32, Fig. 33)

All identical energy orbitums one below other together form **1 s orbit and 3 p orbits.** (Fig.34)

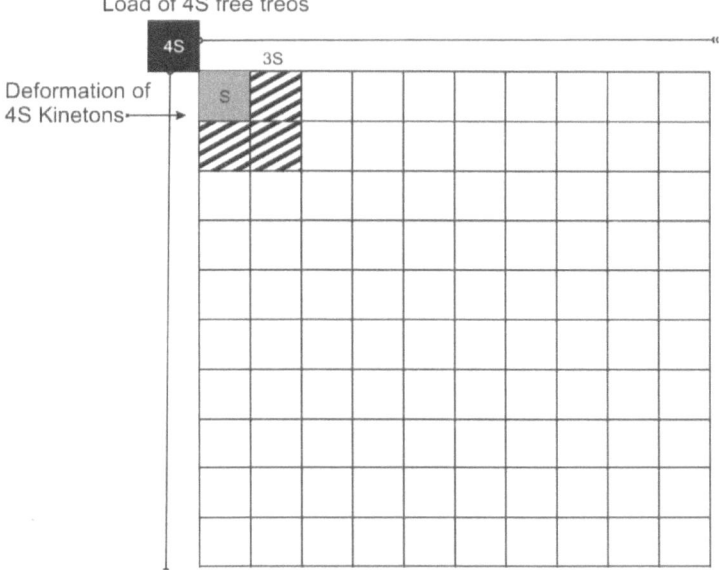

Figure 33: Coloumn geometry in Second dimension at second quantum level (**Top view**)

As, **2 × √S quanta** is mass energy in this two–unit electron packet and in each shell of radius **2 × √S bound treo layers** have 2 sub shells, and 1S + 3 S kinetons from 1+3 squares in two rows, forms 1 + 3 orbitums and thus **2 × √S is frequency of this wave.**

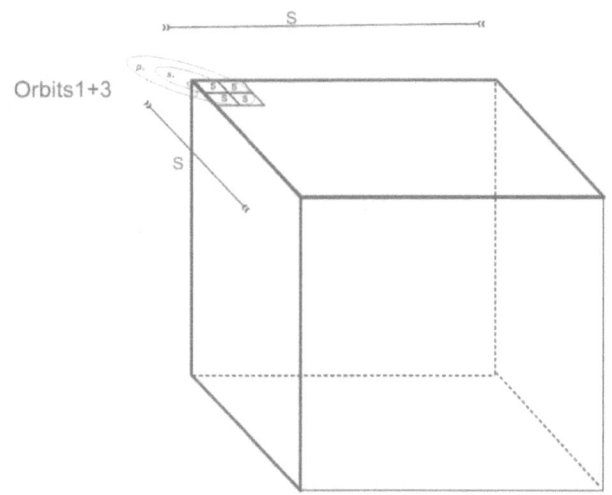

Formation of 1 *S* Orbit by Union of √S Orbitums
and 3 *P* Orbits by Union of √S/2 Orbitums

Figure 34: Coloumn geometry in Second dimension at second quantum level: 1 orbitum from one square and 3 orbitums from 3 squares form one shell. These identical energy orbitums jointly placed, one over other in all shells in wave length, form 1*s* and 3 *p* orbits.

3. Third Quantum Level in Deformation of second Dimension

(addition of new √S bound treo layers and 5 more squares of third row)

The three–unit electron packet is of 3 × √S quanta mass energy or 3 (√S × S free treos). And according to formula S bound treo /3 × √S quanta mass energy in packet and **thus √S/3 apex bound treo is its R C wave length.**

This packet spreads in √S/3 apex bound treo wave length and exert 9 quanta i.e. **3² quanta** load on each apex bound treo.

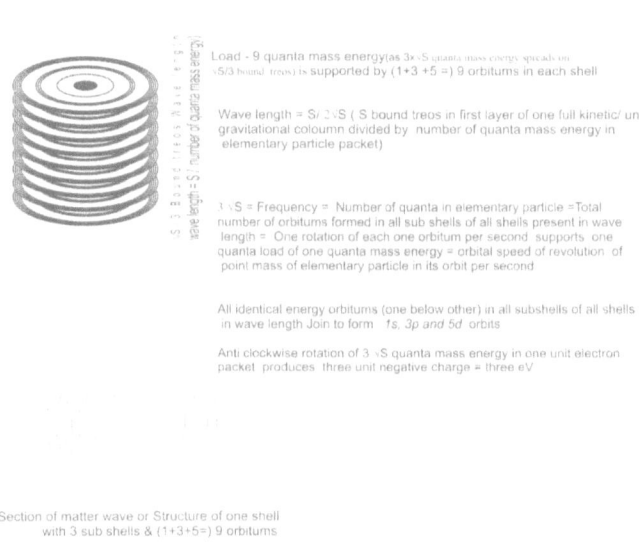

3 'unit electron packet' with 3x √S quanta mass energy at third quantum level in second dimensional deformation of unit space matrix in depth

Load - 9 quanta mass energy(as 3x √S quanta mass energy spreads on √S/3 bound treos) is supported by (1+3 +5 =) 9 orbitums in each shell

Wave length = S/ 2 √S (S bound treos in first layer of one full kinetic/ uni gravitational coloumn divided by number of quanta mass energy in elementary particle packet)

3 √S = Frequency = Number of quanta in elementary particle = Total number of orbitums formed in all sub shells of all shells present in wave length = One rotation of each one orbitum per second supports one quanta load of one quanta mass energy = orbital speed of revolution of point mass of elementary particle in its orbit per second

All identical energy orbitums (one below other) in all subshells of all shells in wave length Join to form 1s, 3p and 5d orbits

Anti clockwise rotation of 3 √S quanta mass energy in one unit electron packet produces three unit negative charge = three eV

Section of matter wave or Structure of one shell with 3 sub shells & (1+3+5=) 9 orbitums

diagramatic representation

Figure 35: Coloumn geometry in Second dimension at third quantum level

With increasing mass energy and decreasing wave length 'Three-unit electron mass energy packet' exert load of 9 quanta at each apex bound treo, which is supported by 1+3+5 orbitums in three sub shells of each shell. All identical energy orbitums one below other in wave length form 1s, 3p and 5 d orbits.

While **3 × √S quanta is mass energy** of this three–unit electron packet and **3 × √S bound treo layers** are deformed from three rows and thus **3 × √S is frequency** of this wave.

1 + 3 + 5 squares are in these first, second and third rows, deforms 1S kineton area + 3 S kineton area + 5 S kineton area = 9 S kinetons area in each shell (Fig. 37).

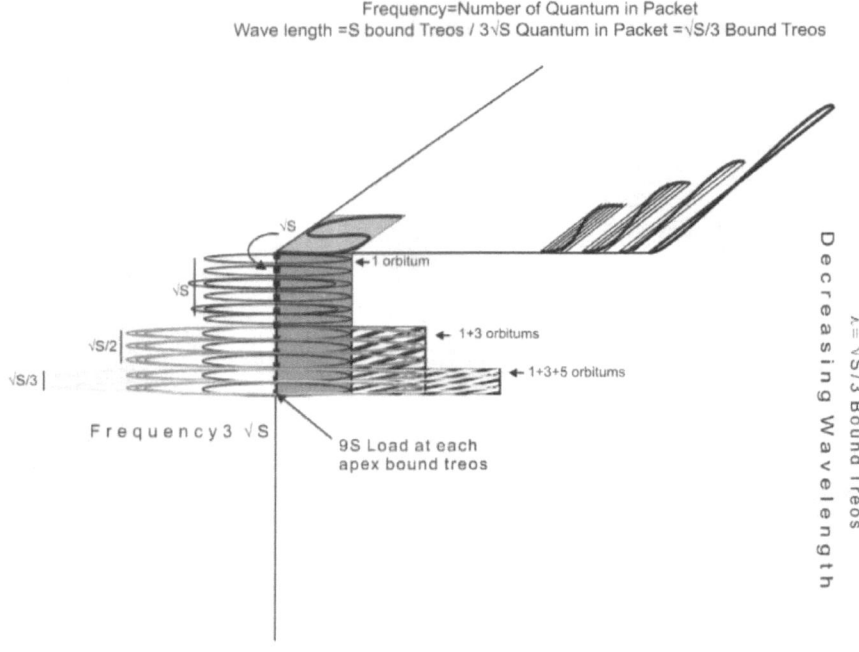

Frequency=Number of Quantum in Packet
Wave length =S bound Treos / 3√S Quantum in Packet =√S/3 Bound Treos

Third Quantum Level Second Dimension

(Diagrammatic Representation)

Figure 36: Coloumn geometry in Second dimension at third quantum level (**side view**): 1 orbitum forms from S kinetons in 1 square of first row, 3 orbitums from 3 squares in second row, and 5 orbitums from 5 squares in third row deforms. For all orbitums in wave length such 1+3+5 squares deform in three rows, in all √S/3 sheets (which deform below second quantum level).

1 square in first row (S kinetons), 3 squares in second row (3S kinetons), and 5 squares in third row (5S kinetons) are deformed in each of √S/3 sheets of unit space matrix just below the deformation of second quantum level. (Fig.36 Fig.37)

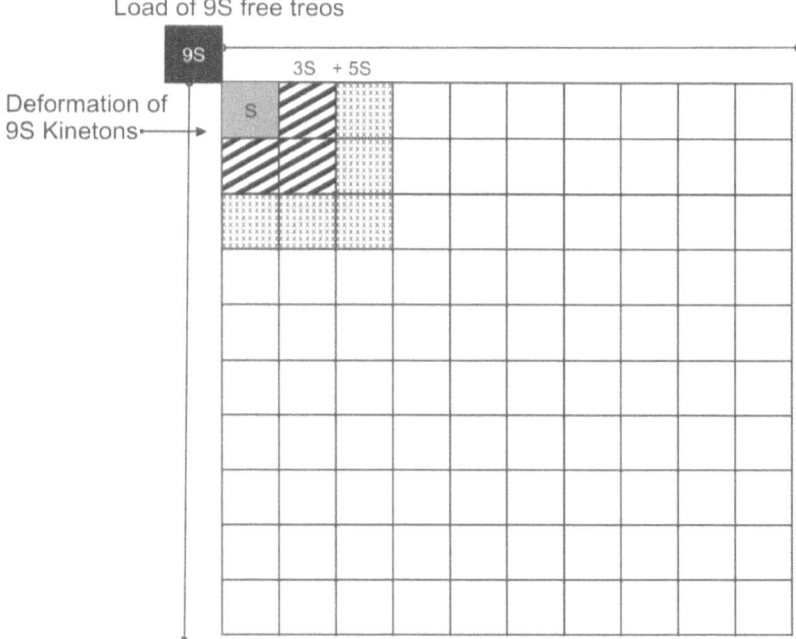

Figure 37: Coloumn geometry in Second dimension at third quantum level (**Top view**)

At Third quantum level which form just below second quantum level of second dimension, each shell has (S + 3S + 5S kinetons) from 1 + 3 + 5 squares in three rows. S kinetons are in one square (= √S × √S kinetons) of first row, 3 S kineton in three squares of second row and 5S kinetons in 5 squares of third row which form (1+3+5 orbitums) in three sub shells in each shell present at √S/3 apex bound treos in vertical wave length. Below the deformation at first and second quantum level; it deforms √S/3 sheets in corner of unit space matrix which gets involved to form √S/3 shells. (Fig. 36)

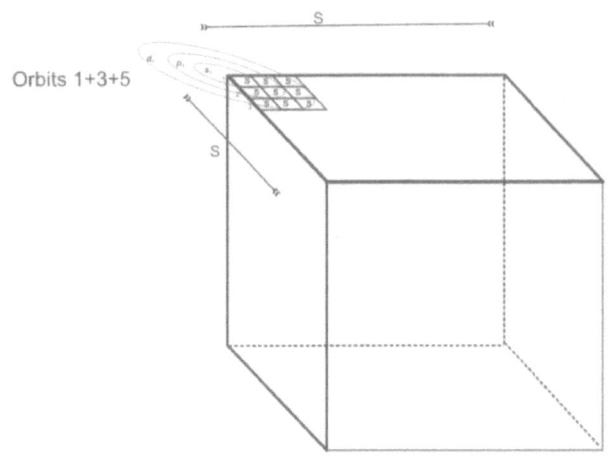

Formation of 1 *S* Orbit by Union of √S Orbitums.
3 *p* Orbits by Union of √S/2 Orbitums
5 *d* Orbits by Union of √S/3 Orbitums

Figure 38: Coloumn geometry in Second dimension at third
quantum level: 1 orbitum from one square in first row and 3 orbitums from 3 squares
in second row, 5 orbitums from 5 squares in third row are formed. These identical
energy orbitums jointly placed, one over other in all shells in
wave length, form **1s, 3 p, 5 d orbits.**

As it forms 1, 3 and 5 orbitums in three sub shells in each shell present
at each apex bound treos in wave length. All identical energy orbitums
one below other in wave length together form *1s, 3p and 5d* **orbits.**
(Fig. 38)

**4. Fourth Quantum level in deformation of second dimension
(addition of new √S bound treo layers and 7 squares)**

The Four–unit Electron packet is of 4 × √S quanta mass energy or 4
(√S × S free treos). And according to formula S bound treo/4 √S quanta
mass energy in packet and √S/4 apex bound treo is its vertical **R C
wave length.** (Fig. 39 and 40)

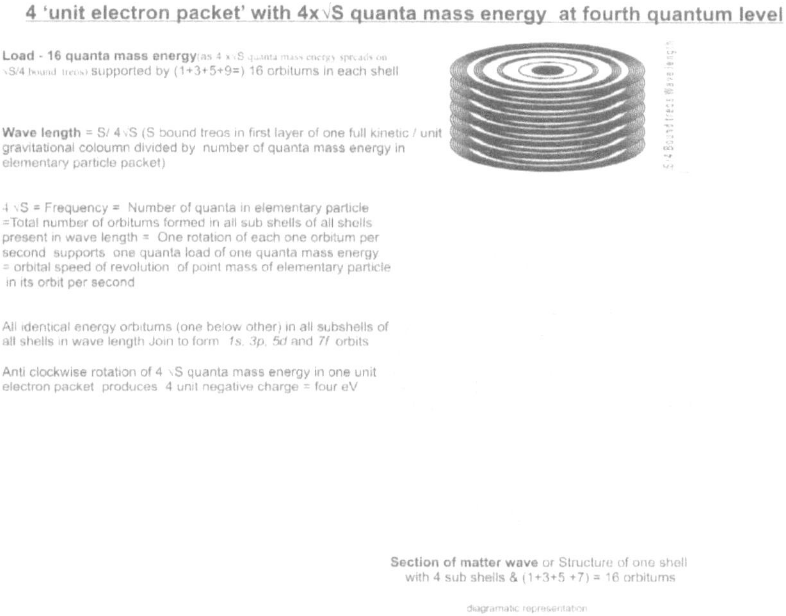

4 'unit electron packet' with 4x √S quanta mass energy at fourth quantum level

Load - 16 quanta mass energy(as 4 x √S quanta mass energy spreads on √S/4 bound treos) supported by (1+3+5+9=) 16 orbitums in each shell

Wave length = S/ 4 √S (S bound treos in first layer of one full kinetic / unit gravitational coloumn divided by number of quanta mass energy in elementary particle packet)

4 √S = Frequency = Number of quanta in elementary particle =Total number of orbitums formed in all sub shells of all shells present in wave length = One rotation of each one orbitum per second supports one quanta load of one quanta mass energy = orbital speed of revolution of point mass of elementary particle in its orbit per second

All identical energy orbitums (one below other) in all subshells of all shells in wave length Join to form 1s, 3p, 5d and 7f orbits

Anti clockwise rotation of 4 √S quanta mass energy in one unit electron packet produces 4 unit negative charge = four eV

Section of matter wave or Structure of one shell with 4 sub shells & (1+3+5 +7) = 16 orbitums

diagramatic representation

Figure 39: Coloumn geometry in Second dimension at fourth quantum level.

With increasing mass energy and decreasing wave length 'Four–unit electron mass energy packet' exert load of 16 quanta at each apex bound treo along its √S/4 wave length. 16 quanta i.e. 4^2 **quanta** load is supported by each shell present at all apex bound treo in wave length.

Each shell has 4 sub shells in breadth which comes from 4 rows. Each row is of √S bound treo layers which form one sub shell. They have 1 +3 +5 +7 squares in four rows, and with 1S kineton area, + 3 S kineton area, + 5 S kineton area and + 7 S kineton area = 16 S kinetons area is of each shell in which it forms 1 + 3 + 5 +7 = 16 orbitums. Each orbitum of S kineton supports S free treo (one quantum mass energy) of packet by its one rotation in one second. (Fig 39)

Thus **4 × √S quanta is mass energy** of packet and **4 × √S bound treo layers are deformed in four rows** thus **4 × √S is frequency** of this wave.

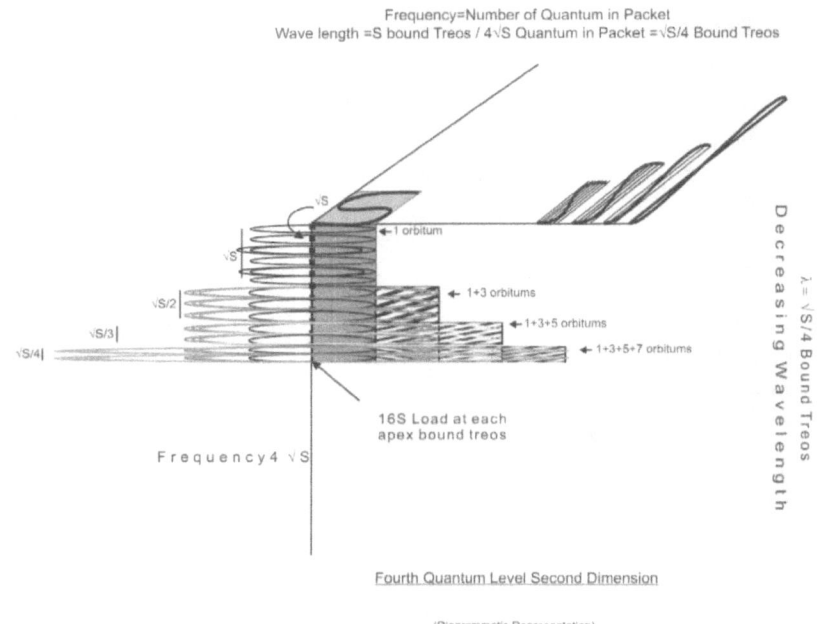

Frequency=Number of Quantum in Packet
Wave length =S bound Treos / 4√S Quantum in Packet = √S/4 Bound Treos

Figure 40: Coloumn geometry in Second dimension at fourth quantum level (**side view**) 1 orbitum forms from S kinetons in 1 square of first row, 3 orbitums from 3 squares in second row, 5 orbitums from 5 squares in third row and 7 orbitums from 7 squares in third row forms. For all orbitums in each shell, such 1 + 3 + 5 + 7 squares in four rows, are deformed in all √S/4 sheets (which deform below third quantum level).

1 square in first row (S kinetons), 3 squares in second row (3S kinetons), 5 squares in third row (5S kinetons) and 7 squares from fourth row (7S kinetons) are deformed in each of √S/4 sheets of unit space matrix just below the deformation produced at third quantum level. (Fig. 40 and 41)

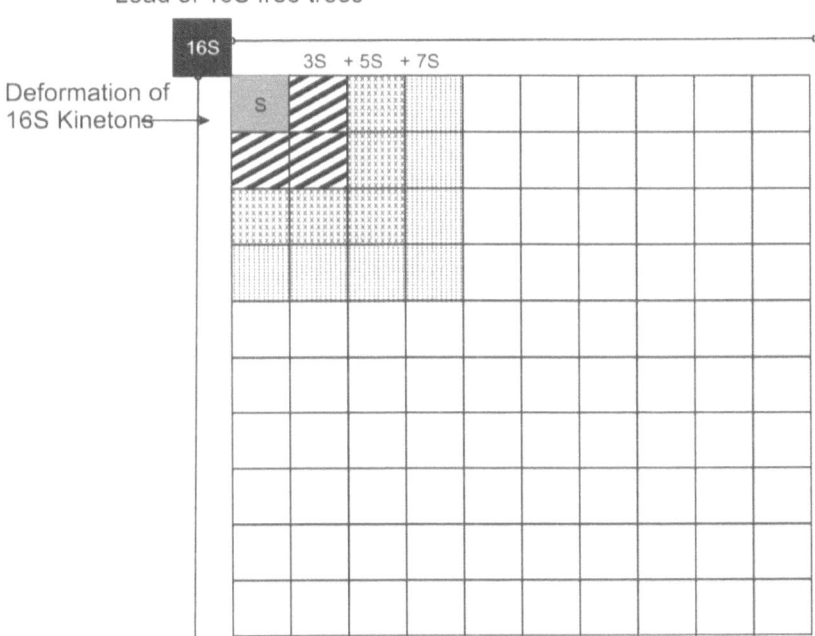

Figure 41: Coloumn geometry in Second dimension at fourth quantum
level (Top view)

Thus, one orbitum of S kineton comes from one square in first row, 3 S
kinetons and 3 orbitums comes from three squares in second row, 5 S
kinetons and 5 orbitums from 5 squares in third row and 7 S kinetons
and 7 orbitums from 7 squares in fourth row, are in 4 sub shells of each
shell present at all √S/4 apex bound treos in wave length.

All identical energy (1 + 3 + 5 + 7) orbitums one below other jointly
forms *1s, 3p, 5d and 7f* orbits (Fig.42).

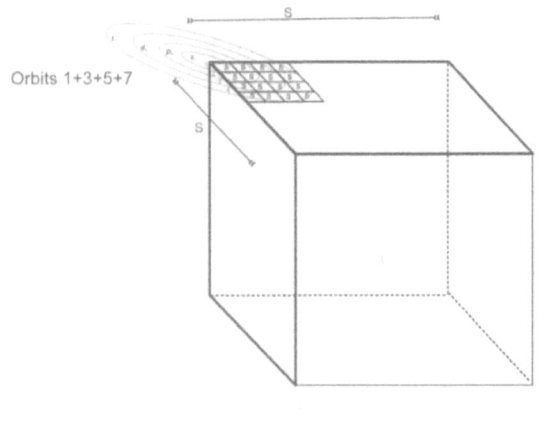

Orbits 1+3+5+7

Formation of 1 S Orbit by Union of √S Orbitums.
3 p Orbits by Union of √S/2 Orbitums
5 d Orbits by Union of √S/3 Orbitums
7 f Orbits by Union of √S/4 Orbitums

Figure 42: Coloumn geometry in Second dimension at fourth quantum level: 1 orbitum from one square in first row, 3 orbitums from 3 squares in second row,5 orbitums from 5 squares in third row and 7 orbitums from 7 squares in fourth row are formed. These identical energy orbitums jointly placed, one over other in all shells in wave length, form 1s, 3 p, 5 d, 7 f orbits.

5. Last or √S Quantum Level in Deformation of Second Dimension

The √S units of unit electron mass energy packet accumulates as, √S × √S quanta mass energy; or S quanta mass energy (or S^2 free treos), which are in **one unit mass** and it is also one Planck's mass $(2.16 \times 10^{-8}$ kg$)$.

According to formula S bound treo/S quanta mass energy in packet; thus, only **one apex bound treo is RC wave length of one unit mass.**

But as **S quanta is mass energy** in this one–unit mass. Now with the involvement of √S rows diagonally and each of √S bound treo layers and up to √S quantum levels, it is total **S bound treo layers** (= √S quantum levels × √S bound treo layers which increases at each quantum level) are deformed to form this shell and thus **S is frequency** of this wave.

1 square in first row (S kinetons), 3 squares in second row (3S kinetons), 5 squares in third row (5S kinetons) and 7 squares from fourth row (7S kinetons) and so on 2n −1 squares in any n th row ....and then finally 2√S − 1 squares are added in last √S^th row and thus it involve one full last sheet of unit space matrix (in depth), at last quantum level. This last one sheet has S quanta or S^2 kinetons, in total S number of small squares which are present in all √S rows (Fig. 43)

From √S rows in last full sheet, total √S × √S = S squares are deformed, which have total S quanta kinetons (S orbitum form each of one quanta) which form one full matter wave in last layer to support **one–unit mass** of S quanta mass energy or S^2 free treos.

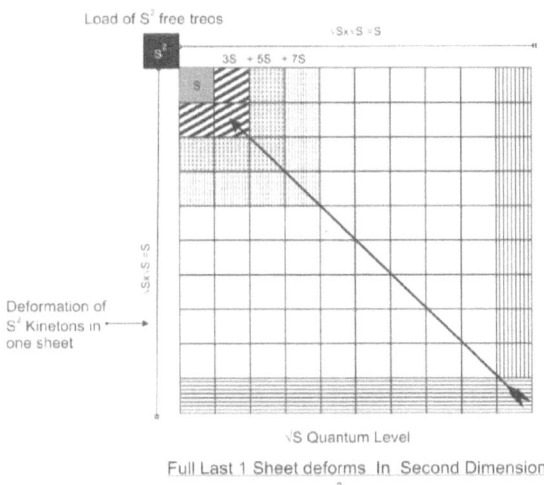

Figure 43: Coloumn geometry in Second dimension at last √S quantum level (**Top view**): 1 square + 3 squares + 5 squares + 7 squares and so on ... 2n −1 squares in any n th row ....and then finally 2√S − 1 squares are added inlast √S th row; thus it finally involve one full last sheet of unit space matrix in depth, at last quantum level, with its one wave length.

Thus, a matter wave which form at last quantum level is of 1 bound treo RC wave length i.e. at only one apex bound treo, or at the unit gravitational center (**at one graviton**) and it supports one–unit mass.

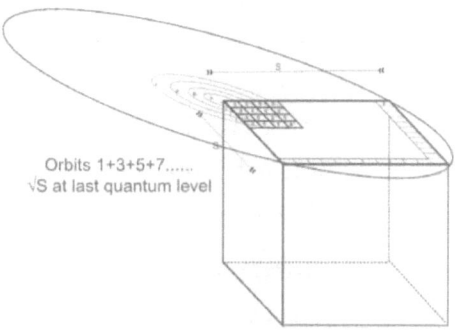

Orbits 1+3+5+7......
√S at last quantum level

Formation of 1 *S* Orbit by Union of √S Orbitums.
3 *p* Orbits by Union of √S/2 Orbitums
5 *d* Orbits by Union of √S/3 Orbitums
7 *f* Orbits by Union of √S/4 Orbitums
2√S-1 orbits at last quantum level each of 1 orbitum

Figure 44: Coloumn geometry in Second dimension at last √S quantum level: S orbitums in last one sheet and total S orbitums at all quantum levels togather in one graviton coloumn, support one–unit mass of S^2 free treos.

The last matter wave is formed by one revolving shell, which have √S sub shells (in last sub shell it adds 2√S–1 orbitums) and total S orbitums or S orbits, with total involvement of S bound treo layers in length and S bound treo layers in breadth in last sheet in total full deformation of second dimension (Fig. 44).

8. Deformation in Three Dimensions of Length, Breadth and Depth

(To support from one unit mass to *√S* unit masses body)

Unit mass is the maximum 'load' (approximately equal to one flea egg), which is supported at its one–unit gravitational center, at one graviton placed **at one bound treo of space matrix**. But then how bigger masses like cosmic bodies and galaxies are supported by space matrix?

Maximum load of **√S unit masses body (of 10^{13} kg** or one billion metric ton)[18], which exert the 'load' of S unit masses **(square of number of unit masses in body) at its gravitational center,** is supported by equal number of S gravitons. All these S gravitons are placed according to column geometry (2n-1 in any n^{th} layer and n^2 in coloumn), in one 'kinetic coloumn of third dimension' of √S spiral graviton layers, named as **one 'electron black hole' (named as it is of size of one unit electron).**

Theme of Deformation in third dimension – In universe, @ **one free treo is supported by one kineton** and thus each unit mass (S^2 free treos) is supported at one graviton by its one graviton coloumn made up of S^2 kinetons.

But **from one–unit mass to √S unit mass bodies** which exerts square number of unit masses load at its gravitational centre, is supported by square number of one to S gravitons in increasing number of spiral layers (from one to √S spiral layers) at 1 to √S quantum levels, in formation of one electron black hole; which form one cyclonic wave in third dimension.

For the biggest √S unit mass body supported in third dimension (i.e. of approx. 10^{13} Kg; one billion metric ton), one kinetic coloumn of electron black hole of √S spiral concentric layers directs a combined kinetic pressure (of all S gravitons present in this kinetic coloumn) towards its apex at the gravitational center of body, to support its S unit

---

18 (For even bigger cosmic bodies of more than √S unit masses the deformation involves fourth dimension of time and then around its unit gravitational center one gravitational sphere forms to support the load of square number of unit masses in body by equal number of gravitons in this gravitational sphere, with the deformation of all four dimensions of Space-Time.)

mass load. (As per coloumn geometry, 2n−1 gravitons are present in any n^{th} spiral layer of this one electron black hole while n^2 gravitons are in n layered kinetic coloumn)

The Electron black hole is one full kinetic coloumn of third dimension having √S spiral concentric layers (which get compacted to the size of one electron of 10^{-13} meter), and has total S gravitons formed by deformation of S^3 bound treos (all pages of one book) in complete three−dimensional contraction of one cube of one−unit space matrix.

The gravitational field of this electron black hole will form by union of graviton coloumns of 2√S −1 gravitons present in outermost layer of this electron black hole.

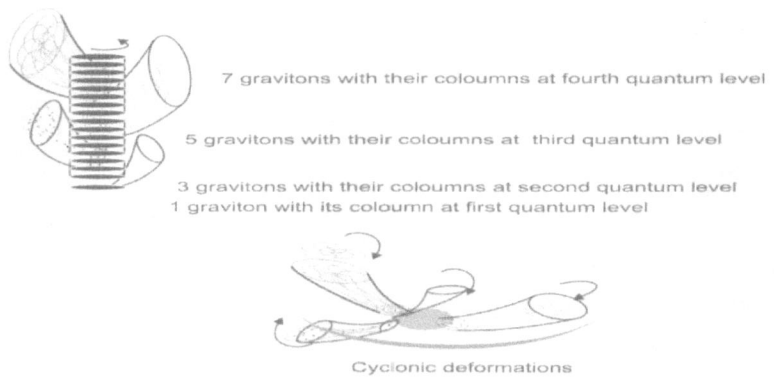

7 gravitons with their coloumns at fourth quantum level

5 gravitons with their coloumns at third quantum level

3 gravitons with their coloumns at second quantum level
1 graviton with its coloumn at first quantum level

Cyclonic deformations

DEFORMATION IN THIRD DIMENSION

One, three, five, seven and so on.... up to a maximum of √S full sheats of 'one unit matrix' deforms (each as one full kinetic/ unit gravitational coloumn of S^2 kinetons) and wrap along with their equal number of unit gravitational centers' (gravitons) at √S quantum levels, in the deformation of third dimension of breadth and thus 'one gravitational coloumn' of body in periphery and 'one electron black hole' forms at center. We observe this phenomenon in nature as seen in sattelite pictures of wrapping of kinetic coloumns at 'eye of an cyclone'.

Figure 45: Coloumn geometry in complete deformation of Third dimension

We can actually visualize such bigger deformations of third dimension; in satellite picture of 'eye of a cyclone' along with circulating and winding layers of graviton coloumns; which produce havoc in cyclone effected areas. (Fig.45 and fig.46)

(a) First Quantum level in deformation of third dimension

One deformed last full sheet (last page of our book for understanding) of unit space matrix in second dimension, with its S^2 kinetons converted as one graviton coloumn, wraps at first quantum level of third dimension and it support one–unit mass at one graviton in 1st spiral layer of electron black hole. (fig. 47a)

CENTER OF A CYCLONE
SHOWING ITS 'EYE'

(example of third dimensional deformation of unit space matrix)

Figure 46: Coloumn geometry in Third dimension as seen in nature

(b) Second Quantum level in deformation of third dimension

Three deformed full sheets (3 pages above last page) of unit space matrix just above first quantum level each with S^2 kinetons (as one graviton coloumn), wraps at second quantum level to support two–unit masses in a body, with its exerted load of 4 unit masses, {in square of unit masses i.e. of $(2)^2$ unit masses at its gravitational center}, by 4 gravitons = 1 + 3 gravitons, in 1st and 2nd spiral layer, where 1 + 3 full sheets (each forming its one graviton coloumns) of unit space matrix are wrapped, while making an two layered kinetic coloumn of electron black hole. (Fig. 47 b)

(c) Third Quantum level in deformation of third dimension

Five deformed full sheets; 5 pages, above pages already deformed (1+3 pages) of unit space matrix just above second quantum level, each having S^2 kinetons (and each forming one graviton coloumn) wraps at third quantum level to support three–unit masses in a body with its load of $(3)^2$ unit masses at 9 gravitons = 1 + 3 + 5 gravitons, in 1st, 2nd and 3rd spiral layer of electron black hole (Fig. 47c)

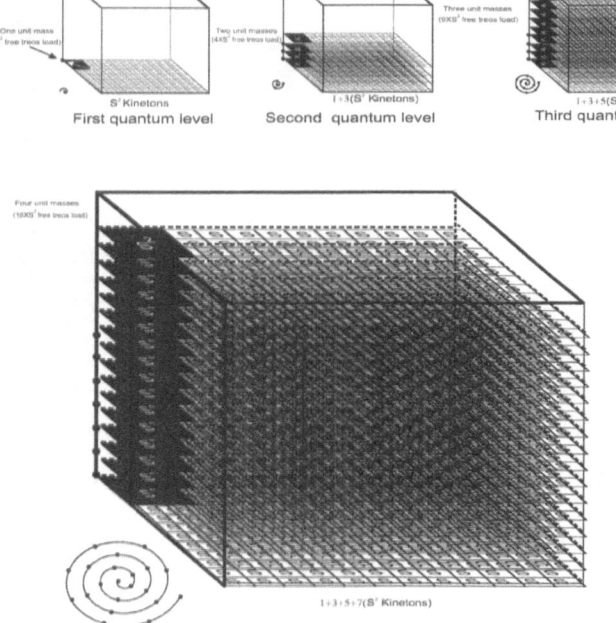

Fourth quantum level
Formation of electron black hole in three dimensional deformation

Figure 47: Coloumn geometry in Third dimension at 1st, 2nd, 3rd and 4th quantum level.

(d) Fourth Quantum level in deformation of third dimension

Now at fourth quantum level 7 gravitons and 7 graviton coloumns, which are from **7 full sheets** [7 pages above, pages already deformed (1+3+5 pages)], adds in fourth spiral layer in deformation at fourth quantum level of third dimension, to support four–unit mass in a body with its total load of 16 unit masses which is supported by 16 gravitons = at 1 + 3 + 5 +7 gravitons in 1st, 2nd, 3rd and 4th spiral layer of electron black hole. (Fig. 47)

With addition of one-unit mass at any n^{th} quantum level, 2n−1 new graviton with 2n−1 graviton coloumns from 2n−1 sheet wrap to form any n^{th} spiral layer of electron black hole.

(e) Last √S Quantum level in deformation of third dimension

In last spiral layer, 2√S−1 gravitons and 2√S−1 graviton coloumns from 2√S−1 sheets add in last $√S^{th}$ spiral layer of electron black hole at √S quantum level of third dimension.

When 2 √S −1 gravitons are added in last spiral layer, then this √S layered kinetic coloumn, supports total √S unit masses in a body with its load in square of unit masses i.e. of $(√S)^2$ or S unit masses by total S gravitons present at all quantum levels of electron black hole. (Fig. 48)

Thus √S unit masses body (or approximately 10^{13} Kg mass in a body i.e. one billion metric ton) is supported by S gravitons in its √S spiral layers in one √S layered 'electron black hole'

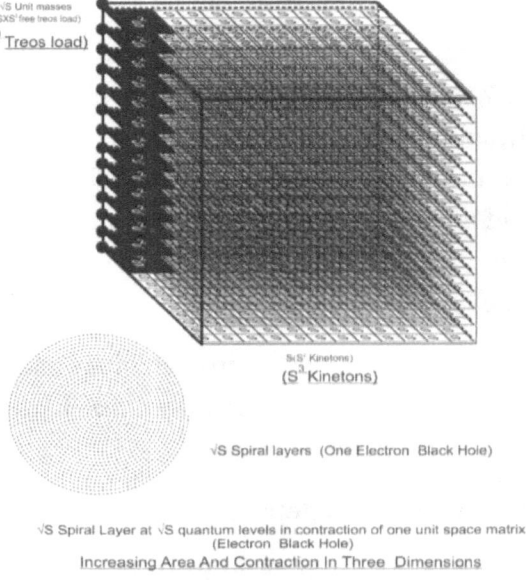

√S Spiral Layer at √S quantum levels in contraction of one unit space matrix
(Electron Black Hole)
Increasing Area And Contraction In Three Dimensions

Figure 48: Coloumn geometry in Third dimension at last quantum level.

9. Deformation in Four Dimensions of Space-Time

(From bodies of one billion metric ton to all big cosmic bodies, and finally up to one unit black hole).

In the deformation of fourth dimension at each of √S quantum levels, √S unit masses increase as one unit, along with increase of √S bound treo layers at each next quantum level in gravitational sphere of body.

At last in biggest gravitational sphere the biggest body of S (or 10^{43}) unit masses exert its load (in all directions on space matrix) of $(10^{43})^2$ unit masses which is perceived and supported at its gravitational center from all directions by equal number of $(10^{43})^2$ rotating gravitons present in gravitational sphere of this **one–unit black hole**. Thus, it deforms total S bound treo layers each having $2n-1$ gravitons in its each nth layer, while total S^2 gravitons are in this biggest possible gravitational sphere of S layers.

This S layered gravitational sphere of unit black hole, having total S gravitons and wrapped S graviton coloumns is formed by contraction of S number of unit space matrices or S **number of full books** (Table 4, in last of chapter 6).

Such multiple unit black holes, increasing as one unit black hole at each of $\sqrt{S}$ quantum level in total deformation of space - time, along with gradual slow down of time from $\sqrt{S}$ vibrations (at gravitational centre of one unit black hole) to zero vibration per second at Galactic center will form different increasing size of galaxies, and each unit black hole will form one spiral arm of Galaxy representing its individual gravitational field.

Basis of Deformation @One free treo is supported by one kineton; In fourth dimension $\sqrt{S}$ unit masses (as one unit) are added, one by one at each of $\sqrt{S}$ quantum levels, with the deformation of additional $\sqrt{S}$ bound treo layers at each next quantum level (it is repetition of pattern of second dimensional deformation), and incorporates $2n-1$ electron black holes (each of $\sqrt{S}$ bound treo layers) at each nth quantum level (e.g. 1+3+5 electron black holes are added at first second and third quantum level; where one electron black hole form by deformation of one cube of one unit space matrix (one book).

Thus, total S number of electron black holes (or S^2 gravitons) are in biggest gravitational sphere of S bound treo layers which support S unit masses of one–unit black hole. (Fig. 49)

The load at gravitational center of one–unit black hole made up of S unit masses, is S^2 unit masses load or of S^4 free treos (S^2 unit masses × S^2 free treos in each unit mass) which is supported at this gravitational center by S^4 kinetons (S^2 kinetons are in one graviton coloumn present on each graviton × S^2 gravitons) which are present in total S number of electron black holes in this gravitational sphere of S bound treo layers.

Biggest Gravitational sphere (Unit Black Hole)

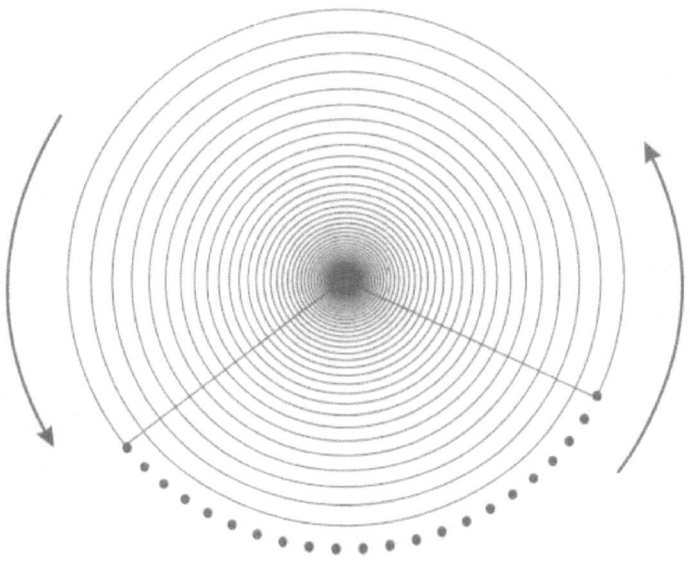

2 S- 1 Gravitons in last bound treo layer
(in 2 √S- 1 electron black holes at Last Quantum Level)

Increasing Area And Contraction of S bound treo layers
(√S bound treo layers at each of √S quantum levels)

Increasing Area And Contraction In Four Dimensions of Space-Time

Figure 49: Coloumn geometry of unit black hole at last quantum
level of 4th dimention.

At each of √S quantum levels additional √S bound treo layers deforms
and with increasing radius of gravitational sphere it has one, three,
five and seven electron black holes at first, second, third and fourth
quantum level and so on 2n−1 electron black holes are added at any n^{th}
quantum level, alongwith new √S bound treo layers.

In last, at √S quantum level 2√S−1 electron black holes and 2S−1
gravitons along with its graviton coloumns are added in lastly added
S^{th} layers, in this biggest S layered gravitational sphere of one−unit

black hole. Thus, one S unit masses body is supported by S^2 gravitons (present in S electron black holes) present in S layered gravitational sphere or one– unit black hole.

In deformation of fourth dimension of Time, in S bound treo layers of biggest gravitational sphere of unit black hole; from S vibrations per second **(of all kinetons)** at its peripheral layer the **number of vibrations is curtailed @ one vibration per gravitational coloumn layer per quantum level** towards its center. Thus with reduction of vibrations one by one in each concentric gravitational coloumn layer towards centre of S layered gravitational sphere, the time will thus slow down at gravitational center of one unit black hole.[19]

---

19 One solar mass, M , can be converted to related units:

27068510 ML (Lunar mass)

332946 M (Earth mass)

1047.56 MJ (Jupiter mass)

1988.55 yotta tonnes

It is also frequently useful in general relativity to express mass in units of length or time.

- M $G/c^2 \approx 1.48$ km (half the Schwarzschild radius of the Sun)
- M $G/c^3 \approx 4.93$ μs

Explanation of MG/c^2

(!) (a) According to treo model when the value of G is derived in Planck's units it is 10^{94} kinetons per Kg per second per second × M i.e. 10^{30} Kg as mass of Sun, and then MG calculates mass energy of Sun in treos MG = (M10^{30}Kg×G10^{94}) number of treos (=10^{124}) and then MG/c^2 (× $10^{94}/10^{86}$) converts it in number of unit masses in body of Sun 0.91379754 × 10^{38} unit masses

(b) It is minimum length of space matrix in shape of gravitational sphere of Sun which can support Sun and It can be expressed in length as 0.91379754×10^{38} Bound treo layers or 1.48 Km or radius of Schwarzschild sphere of Sun (the same is calculated by general theory of gravitation by Einstein).....

(!!) One-unit mass is the maximum load which can be supported at one point (at one bound treo) at one Graviton by space matrix. $(0.91379754×10^{38})^2$ unit masses is load at gravitational center of 0.91379754×10^{38} unit masses in body of sun, which is supported by 0.91379754×10^{38} Bound treo layers in gravitational sphere of Sun by $(0.91379754×10^{38})^2$ gravitons.

Smaller size cosmic body like our Sun is made up of 10^{38} unit masses, which exert a load of $(10^{38})^2$ unit masses at its gravitational center. To support the gravitational center of Sun, a 10^{38} bound treo layered gravitational sphere is formed (**it is the same size as calculated by general theory by Einstein**), which have $(10^{38})^2$ gravitons (@ $2n-1$ gravitons are in each nth bound treo layer and n^2 gravitons are in full gravitational sphere) as per coloumn geometry. (Fig. 50)

All rotating graviton coloumns of these gravitons in all layers of gravitational sphere, including **$2 \times 10^{38}-1$ graviton coloumns at outermost peripheral layer merge together to form gravitational field of Sun.**

Gravitational sphere of Sun

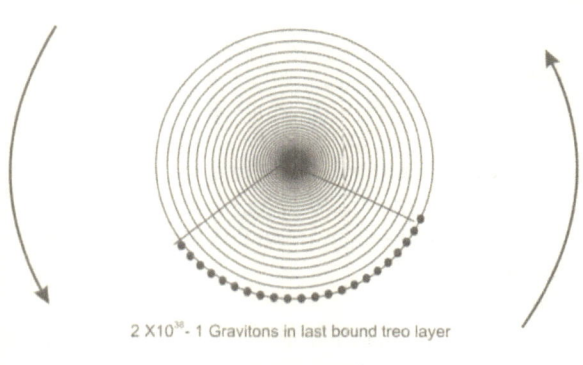

2 X10³⁸- 1 Gravitons in last bound treo layer

Increasing Area And Contraction of 10^{38} bound treo layers
in Gravitational sphere of Sun

Increasing Area And Contraction In Four Dimensions of Space-Time

Figure 50: 10^{38} unit masses in body of Sun at its gravitational center is supported by
10^{38} bound treo layered gravitational sphere (of 3 Km diameter)
having $(10^{38})^2$ gravitons.

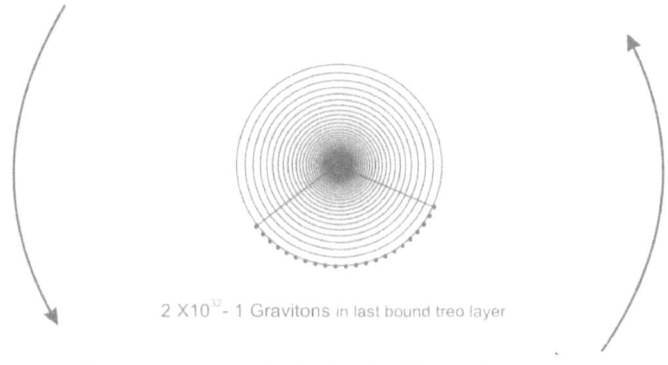

Gravitational sphere of Earth

2 X10¹²- 1 Gravitons in last bound treo layer

Increasing Area And Contraction of 10¹² bound treo layers
in Gravitational sphere of Earth

Increasing Area And Contraction In Four Dimensions of Space-Time

Figure 51: 10^{32} unit masses in body of Earth at its gravitational center is supported by 10^{32} bound treo layered gravitational sphere (of pin head size) having $(10^{32})^2$ gravitons.

Similarly, Earth is made up of 10^{32} unit masses has 10^{32} layered (Pin head size) gravitational sphere. (Fig. 51)

10. Wave Length and Frequency in Third and Fourth Dimension

Frequancy

Frequency of the wave **was increasing by one unit,** with increase of mass energy by **one quantum,** alongwith **one bound treo layer** (one kineton layer) which increases in all sub kinetic coloumns at each of √S quantum levels in first dimension.

Then the **frequency increased in multiple of √S number in second dimension,** with the increase of √S **quanta in packet** at each of √S quantum levels which resulted in **increase of √S bound treo layers** in radius of each shell, present on each apex bound treos of its RC wave length.

Thus at all **√S quantum levels in first** and **√S quantum levels in second** dimension the **frequency increased from 1 number to S number.**

While, the frequency in third and fourth dimension gradually **decreases by one with the addition of one new layer in kinetic coloumn.**

With increase of one unit mass at each next quantum level in deformation of **third dimension the frequency decreases by one** with addition of each new deformed bound treo (graviton) layer in electron black hole.

While **in fourth dimension the frequency decreases by √S number** with deformation of new √S bound treo layers (with 2n-1 electron black holes in one layer at n^{th} quantum level) in gravitational sphere (with increase of √S unit masses of body at each of √S quantum levels in fourth dimension).

Wave length

In first dimension all packets fromed at √S quantum level **gradually contracts and spreads** and on its **gradually reducing wave lengths.** (S bound treos/number of quanta in packet) for one quanta in one unit photon, to √S quanta of gamma photon it forms one EM wave.

While in second dimension the packets will spread on RC wave length (S bound treos/number of quanta in packet) and will form one wave on its 2 RC wave length. In second dimension the increasing angular momentum (2n-1 x π) will calculate the circumferenc of its orbit; which is also Compton wave length of packet. The packet contracts on reducing wave length.

But **wave length in third and fourth dimension gradually increases; with increasing deformation. With increasing value of n, one new wave will form on increased number of gravitons (2n–1) present in** every new n^{th} layer in kinetic coloumn of **third dimension.**

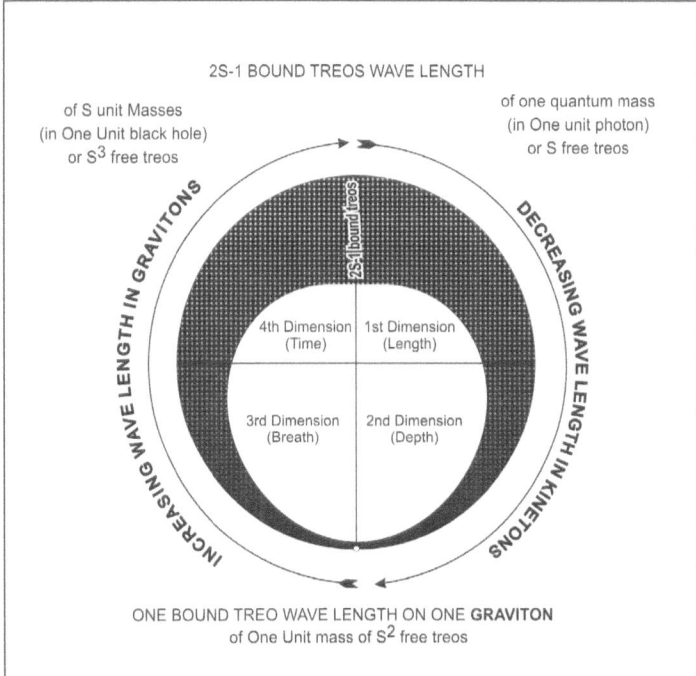

Increasing mass energy from one unit photon (S free treos mass energy) to one unit mass (S^2 free treos mass energy) **gradually contract** on successive **decreasing wave length** from 2S-1 bound treos to just one bound treo wave length (on one **graviton**) with the completion of deformation in first and second dimension.

Then increasing mass energy in unit of 'one unit mass' and its load in 'square of unit masses' at its gravitational center, **'spreads as 'diluted mass pressure'** (as 2n-1 unit masses in any direction) **in surrounding space matrix** on gradually **increasing wave length** as each successive bigger layer from one graviton of 'one unit mass' to 2S -1 gravitons of 'one unit black hole' (S unit masses) and thus form bigger gravitational fields with the completion of deformation in 3rd and 4th dimension.

<u>Diagrammatic representation of spread of load of increasing mass energy in all four dimensions</u>

Figure 52: Mass energy packets contracts on its reducing wave length in first and second dimension. While diluted mass pressure of central load is supported by gravitons in its increasing wave length in third and fourth dimension.

The wave length is 1 graviton in first spiral layer, then 3 gravitons in second spiral layer, 5 gravitons in third, 7 gravitons in fourth and so on...... 2n-1 gravitons in n^{th} layer and at last quantum level it is $2\sqrt{S} -1$ gravitons in last spiral layer.

Similarly **2n−1 electron black holes** are in any one n^{th} layer of kinetic coloumn on which one wave will form in **fourth dimension**.

As there are, 1 electron black holes in first layer, then 3 electron black holes in second layer, 5 electron black holes in third, 7 electron black holes in fourth and so on...... 2n-1 electron black holes in n^{th} layer and at last quantum level it is $2\sqrt{S} -1$ electron black hole in last layer (Each **electron black hole is made up of $\sqrt{S}$ spiral layer**).

So each next quantum level will form only after increase of $\sqrt{S}$ layers of gravitational sphere are deformed with increase in mass.

At each next quantum level the **wave length** will increase by $\sqrt{S}$ gravitons. At any n^{th} quantum level in n^{th} layer one wave will form on $\sqrt{S}$ (2n-1) gravitons, while in last $\sqrt{S}^{th}$ layer on $\sqrt{S}$ $(2\sqrt{S} -1)$ gravitons it will form **one single wave** (on total 2 S gravitons length arc, of a circle of 2S × π circumference, which is also the compton wave length of this body).

(2 × RC wave length x π = Compton wave length) of S radius of one unit black hole.

CHAPTER 4

The World around Us

1. Unit Mass

It is S quanta mass energy (or S^2 Free treos), supported at its one 'unit gravitational center' by S^2 kinetons in one graviton coloumn. **One unit mass is supported** at its unit gravitational centre by apex bound treo of one graviton coloumn or **one graviton.**

In increasing deformation at √S quantum levels of second dimension the mass energy increases succesively **by √S quanta mass** at each next quantum level, and thus finally the biggest **√S layered kinetic coloumn which is now named as graviton coloumn** of second dimension. The biggest mass of S quanta (**as √S quanta mass energy increases at each of √S quantum levels**) is supported by total deformation of second dimensional deformation is this **unit mass** which is *conventional Planck mass 2.176434 × 10⁻⁸ Kg* (the mass roughly equal to size of one flea egg).

One unit mass is *[a miniscule black hole of one planck mass; $(2GM/c^2)$]* is supported at one bound treo or at one graviton, is maximum capacity of space matrix to support a load in universe at one (bound treo) point.

Graviton is a boson with spin 2. Spin two means that by its half rotation the particle will regain its orientation (to understand it, the familier shape is queen of playing cards; by half rotation of this playing card the queen regains its shape)

S^2 free treos are in a unit mass (or *Planck's mass*), are supported by S^2 kinetons with its total energy *Planck's energy of 1.96 × 10⁹J* or **mass energy of one unit mass and the same is kinetic energy of one graviton.**

(A) The energy in one Kg mass when calculated in Joule according to Einstein's equation as E = mc²

(B) If m is one Kg and c² (3 × 10⁸ meter per sec)² = E = 9 × 10¹⁶ J

(Kg × meter ²/second ² ; Dimensional formula of energy in joule; m l ²/t ²)

(C) One Planck mass or unit mass (known value) = 2.176434 × 10⁻⁸ Kg

(D) One Planck mass (energy) = 9 × 10¹⁶ J × 2.17643 × 10⁻⁸ Kg = 1.96 × 10⁹J

(E) As per treo model; one Planck mass (unit mass) = 3.440499 × 10⁸⁶ free treos energy.

(As One Planck mass = *2.176434 × 10⁻⁸ Kg* and 1.58079692⁹⁴ free treos are present in one Kg[20]. Then; *2.176434 × 10⁻⁸* × 1.58079692⁹⁴ = **3.440499 × 10⁸⁶ free treos energy is in one unit mass** = S^2 free treos energy = S quanta energy.)

(F) The energy of one unit mass can also be calculated by equation *E = m c²*

Then calculated energy **E of 1 unit mass** = 1 unit mass × [21](1.85485844× 10⁴³ bound treos distance per second)² = 1 × 3.440499×10⁸⁶ = 3.440499 ×10⁸⁶ free treos (bound treos distance, per second)²

Therefore One unit mass = one Planck mass = 2.176434 × 10⁻⁸ Kg = 3.440499 ×10⁸⁶ free treos = *1.956 × 10⁹J is mass energy of one unit mass.*

---

20   see calculation of gravitational constant page 218
21   (Speed of light = 1.85485844×10⁴³ bound treo distance per second) See page 43

And as GRAVITON has equal number of 3.440499 ×10^{86} kinetons, thus this kinetic energy of one graviton *is also = 1.956 × 10⁹J is kinetic energy of one graviton = **Planck energy.***

(!) [22]**Number of gravitons in one–meter length (one on each bound treo)** = 0.618714×10^{-35} bound treos in 1 meter; when multiplied by kinetic energy of one graviton we get =1.956×10⁹ J × 0.618714×10^{-35} gravitons per meter = known value of **Planck's force of 1.2102×10^{44} Newton** (J per meter)

(!!) **kinetic energy of gravitons in one–meter cube** calculates known value of **Planck's energy density.**

    (a) **5.155 ×10^{96} Kg/m³ × 9 × 10^{16}J = 4.633 × 10^{113}J/m³** (conventional calculation)

    (b) **1.956×10⁹J** × (0.618714 × 10^{-35})³ = **4.633 × 10^{113}J/m³** (calculations as per treo model)

Thus number of bound treos in one meter length = (1/ 1.616229 × 10^{-35}) = 0.618714×10^{-35}

These known values of *Planck's energy, Planck's force* and *Planck's energy density,* finds its justification as energy of accumulated gravitons at one bound treo (*Planck's energy*), at all bound treos in one meter (*Planck's force*), and on all bound treos in one meter cube (*Planck's energy density)* is thus explained and is calculated for the first time.

To support load of bodies made up of multiple unit masses multiple gravitons are required. The gravitons are carrier bosons of gravitational forces and they jointly form kinetic coloumns at gravitational centre (to support increasing number of unit masses load) of bodies named as **electron black hole, gravitational sphers of all cosmic bodies** and **in gravitational sphere of unit black hole** (where Gravitons are in

---

22 As length of one bound treo = 1.616229 × 10^{-35} meter or one Planck least length

3 lac Km radius in the gravitational sphere to support the load of one unit black hole).

2. Charge

a. *1eV, or One–elementary charge of 1.602176634×10^{-19} coulomb; is the charge of one unit electron. (known value)*

b. **But one unit Electron is made up of 1.44×10^{64} free treos. (treo model see page 48)**

c. **Therefore; charge on one free treo is 1.1121 67×10^{-83} coulomb.**

   *(1.602176634×10^{-19} coulomb/***1.44×10^{64}** free treos in one unit electron = **1.11261604 × 10^{-83} coulomb**)

d. **Thus one coulomb charge is on 0.898782656×10^{83} free treos.** (1/1.11261604×10^{-83} = **0.898782656×10^{83} free treos**)

e. *One coulomb charge is on 6.2415 × 10^{18} electrons. (Known value)*

   By calculating it according to treo model we get the same value and it is another proof of accuracy of treo model.

   Number of free treos has *one coulomb* charge/number of free treos in one electron = number of electrons having one *coulomb* charge.

   (0.898782656×10^{83}/1.44 ×10^{64} = **6.2415 ×10^{18} electrons**)

f. *Charge mass ratio = 1.7588×10^{11} coulomb/Kg*

   (1.5808523×10^{94} Free Treos are in one Kg mass (treo model)[23] /**0.898782656×10^{83} free treos have one *coulomb* charge**) By **calculating it according to treo model we get the same value and it is another proof of accuracy of treo model.**

---

23 see calculations of gravitational constant on page 218

RADIATION OF CHARGE ON SPACE MATRIX (AS PER TREO MODEL).

One free treo is the minimum load at first quantum level of first dimension as **S free treos of unit photon** spread on S apex bound treos on its S bound treos wave length, and is supported by one Kineton. **Each free treo** will have **$1.11261604 \times 10^{-83}$ coulomb charge.**

At first quantum level of second dimension, **1 quantum (S free treos)**, is the load at each √S apex bound treos of RC wave length of **unit electron (√S quanta mass energy; have 1 eV charge) and is supported at 1 orbitum.**

The *elementary charge of 1eV is the charge of this one unit electron* i.e. (on √S× S free treos; **or on 1.44×10^{64} free treos**) and it **divides on √S apex bound treos** along its RC wave length, as **charge of S free treos (or of one quanta free treos) at each apex bound treo** at first quantum level of second dimension.

As per 'wave particle duality', the mass energy in any packet can either remain concentrated on all apex bound treo (in wave length), or all free treos in its mass energy can spread @ one free treo on one kineton. Thus one quanta or S free treos will spread on one orbitum made up of S kinetons.

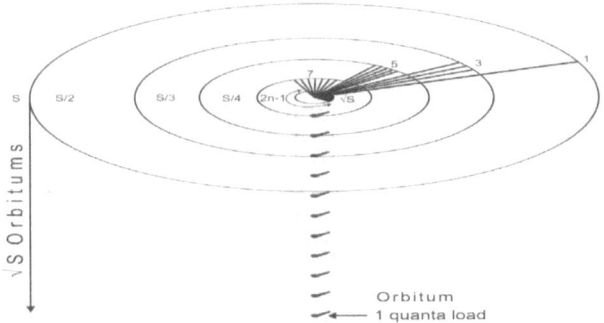

Figure 53. I eV charge on one unit electron radiates as one free treo on one kineton as radiating charge from point source. In two dimensional deformation at first quantum level, one quanta load exerted by unit electron at each apex bound treo, in its RC wave length is supported by one orbitum. In its first dimentioal (contracted) deformation of √S orbitums this 1eV charge of unit electron, spreads.

The direction of rotation of free treos with kinetic coloumns in rotating electrons creates positive and negative charge (isospin and quarks theory which explains charge, is only elobarations of this basic concept described here).

The **motion of free treos** (spread as one free treo on one kineton in kinetic coloumn) along with motion of kinetic coloumns of any advancing charged particle, when seen from its back (receding from you) if they look rotating clockwise (symbolising positive charge), but maintaining the same rotation the same packet when seen from front (approaching you), they will look rotating anti clockwise (symbolising negative charge). Thus the positive and negative charges cant exsist individually (and based on same reason and principal magnets have north and south pole, and one pole is not possible).

The anti clockwise **rotating** free treos with kinetons on rotating layers of shells will exhibit–ve charge and directed radially to wards charge particles.

And clock wise **rotating** free treos with kinetons on these layers will exhibit + charge and directed radially away from charge particles.

The two opposite charges on protons and electrons in an atom produces electro static forces. The magnetic field exert force from perpendicular direction, thus can not change the speed of moving charge particles but can change its direction.

The angular momentum in second dimension devides as Spin angular momentum (responsible for rotation on its axis) and orbital angular momentum (responsible for revolution in its orbit) of particle.

The flow of rotating free treos in units of one electron packet on electric wires is moving charge of 1 eV. While the **supporting kinetons (**magnetrons) in rotating layers of their supporting kinetic coloumns (shells, having sub shells and orbitums), form magnetic field.

The moving energetic electrons on a wire when retarded by a resistance, can shed off *1 eV charge* *of one unit electron* (charge on all 1.44×10^{64} free treos in one electron packet **i.e. √S quanta mass energy); or it may shed of the charge in integral multiple number of free treo quanta (mass energy) along with transfer of kinetic layers on which it is** riding, **from all kinetic coloumn** supporting this electron packet.

This charge as donated free treos mass energy, alongwith its supporting **kineton layers is shifeted from** donating body, to all kinetic coloumns in RC wave length of receiving body. This energy transfer of free treos and charge, with kineton layers as electric power, is used accordingly for its different uses in all electric motors.

With increasing mass energy by one unit electron mass and charge by 1 eV, at each next quantum level **all energetic electrons** are formed **from 1eV to millions of eV** of gradually decreasing wave lengths and increasing electron packet densities. (just like formation of all √ S types of photon packets in deformation of first dimension)

3. Electricity and Magnetism Redefined

The electric field is produced by a scalar source, i.e. electric charge (or number of electrons) which is specified in some units, by **number and their positive or negative charge** (which is decided according the clock wise and anti–clockwise direction of their spin). *While magnetic field is produced by a vector source, i.e. by a current element which has a* **magnitude and direction** *along a line element.*

The electrical field is along the radial vector joining the source and field point, the magnetic field is perpendicular to both the radial vector and to the current element vector.

Thus **radiating magnetic fields are formed along the path of rotating layers of kinetic coloumn of any moving charged particle** or around a moving electron particle in atom or on wire.

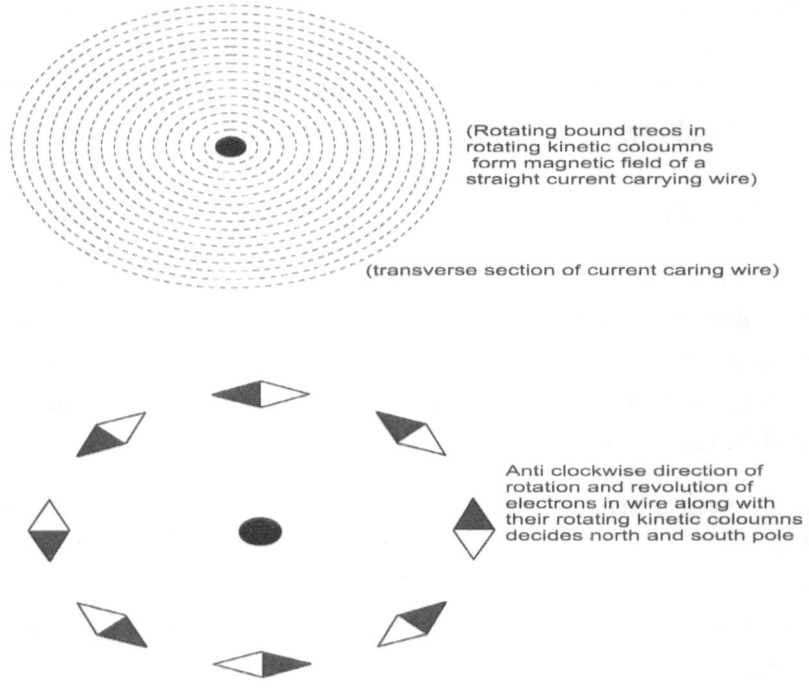

(Rotating bound treos in rotating kinetic coloumns form magnetic field of a straight current carrying wire)

(transverse section of current caring wire)

Anti clockwise direction of rotation and revolution of electrons in wire along with their rotating kinetic coloumns decides north and south pole

Magnetic field produced in long current carrying wire

Figure 54: Rotating kinetons in rotating orbitums of sub shells with each shell present at each apex bound treo in RC wave length of each Electron, which are moving as electric current on current carrying wire, forms magnetic fields.

The magnetism is due to the motion of kinetons in the moving rotating and propelling layers of kinetic coloumns, while pushing a central charge (an electron) along its axis. When you plot magnetic lines of magnetic forces of bar magnet by iron filings, the lines thus plotted on paper are actually the layers of revolving kinetic coloumns.

The magnetism is indirect proof of existence of kinetic coloumns. The rotation of coloumns is not only responsible for spin of electron but also for pushing electron in a linear motion.

Both magnetic and electric field strength reduces inversely by the square of distance, between source and the 'field point'. This type of dispersion of strength of magnetic force indicates that the magnetic fields are arranged according to proposed coloumn geometry.

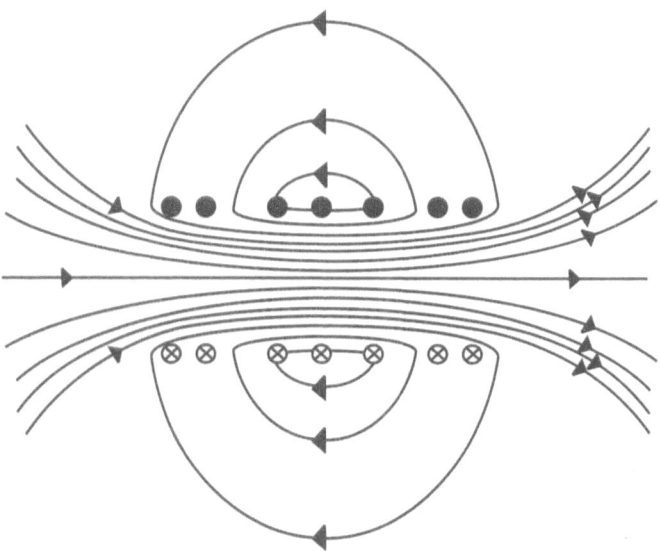

Figure 55: Rotating kinetons in rotating orbitums in sub shells with each shell **at** Quantum levels are depicted by 'magnetic line of forces' in an magnetic field, formed by rotating kinetic coloumns, which supports free electrons circulating in a bar magnet.

The spin of moving electrons and the direction of rotation of the kinetic coloumns, **with the direction of flow of kinetons**, creates two poles of magnet. The kinetons (or magnetrons) in layers of its kinetic coloumns enter from south–pole and when seen from its back, they look rotating clock wise

But when they exit from north–pole, maintaining the same rotation and seen from front, they look rotating anti clockwise.

These lines of forces are always closed loops (unlike electric charge where these lines merge at point charge).

The anti–clockwise revolution of electron, with an external magnetic field perpendicular to the plane of orbit, the direction of its force by Fleming's left hand rule will be outward and for its clockwise orbits the force will be inward.

The density (closeness) of magnetic lines denotes strength of magnetic field, as it denotes increased deformation of space matrix at higher quantum levels. Like all other forces smaller fields merge to produce larger fields, the magnetic fields produced by all electrons or charge particles **merge to produce bigger fields with denser magnetic lines of forces.**

The magnetic strength increases with the increased rate of spin of charged particles. According to treo model increased rate of spin or **the rate of rotation of charge particles** is only possible when the **kineton layers increase** in all supporting kinetic coloumns of these particles, and this increases the magnetic field strength.

The opposite spins of paired electrons and their rotating kinetic coloumns in opposite directions, neutralizes the net magnetic momentum of atom to zero.

Thus, the dia–magnetic substances have no atomic magnetic moment, para–magnetic substances with one valance electron have one bohr magnetron (eh/4 π me = 9.274 x 10^{-24} ampere/ square meter) and ferro–magnetic substances (Fe, Ni, Co) with their incomplete electronic shells have large value of atomic magnetic moment.

Magnetization of a substance means alignment of spin of all the electrons in atomic orbits (in sub shells of each shell) in the direction of legend magnetic field.

(Ref. 8, chapter magnetism '**Inside A Wave**', 2005 Manas Prakashan.)

4. Treo Model Describes Atomic Configrations

In 2005 when one omnipresent and omnipotent field was independantly postulated and documented for the first time by author as Space matrix in his book "INSIDE A WAVE", author was ignorent that some what similar Higgs field was postulated earlier.

But Higgs field were never eloborated, to explain its geometry, properties, purpose and working. While the discription of Space matrix as per this treo model revealed the coloumn geometry, which explained fields of all forces and most of universal pheonomenon and the inferences drawn could be supported by simple mathematical calculations.

In this study of atom in accordance with proposed treo model, **five main problems are explained by coloumn geometry**; (1) Why 2, 8, 18, 32 electrons are placed in atomic orbits, formed at four energy levels with increasing size of shell. (2) How all elements of periodic table are formed by succesive filling of these empty atomic orbits by electrons. (3) Purpose and pattern of distribution of atomic energy in orbits is according to proposed coloumn geometry (4) It suggests one Sinking nucleus model (5) Quantum gravitation at atomic level.

Out of all known 118 elements which are grouped in periodic table, the elements upto iron are formed in stars and for still bigger atoms like gold and uranium (which are created by rapid neutron captures) required very voilent energy fields, provided by exploding stars during its supernova formation or by merger of two neutron stars. ***Only first 98 elements are found naturally on earth and rest were synthetised in nuclear accelerators,*** **but they all could be grouped in periodic table as per coloumn geometry.**

In the deformation at **first quantum level of second dimension**, each shell have only one sub shell and one orbitum (and all orbitums in all these shells togather join to form one orbit).

With increasing mass energy and increasing deformation at each next n^{th} quantum level in all shell, one on each apex bound treo in its RC wave length, one n^{th} Sub shell (e.g. *s*, *p*, *d*, *f* sub shell) is added and it adds thus formed 2n-1 orbits, which are succesively filled by electrons to form all elements, as total atomic deformation occurs at **four atomic quantum levels.**

(a) Deformation And Counter Deformation In Second Dimension.

To visualse the step by step increasing **total deformation of second dimension,** from 1st quantum level with deformation of just 1 square in √S sheeths, to last quantum level with deformation of all S number of squares in one last sheet, we will draw this deformation in a **cube of one unit space matrix** as shown in figure 56 (with its 8 corners A, B, C, D at top and E, F, G, H corners at its bottom, in which E corner is below A corner).

The deformaion strarts from corner A and gradually increases in steps to end after deforming one last (square) sheet with its corners E, F, G, H and thus total deformed area (not drawn in steps) can be marked by joining lines from corner A to E, F, G, H.

The identical Counter-deformaion starts from diogonally opposite corner C and then gradually increasing in steps, this also ends at this last squared sheet (E, F, G, H corners) and this total area deformed can be marked by joining lines from C to E, F, G, H. Now we will notice a point O appears, a meeting point of line AG with line CE, and pyramid shaped area below this point O appears with its base at E, F, G, H, **in which both deformation and counter deformation overlaps,** (See Fig. 56) and while rotating in two opposite directions they forms two vortexes.

(Vortex – is a area where the flow spins around an axis line which can be straight or curved shape. They can complexly move, stretch, twist and inerect with the surroundings.)

The four Quantum levels n1, n2 , n3 , n4 of deformation and same quantum levels are from counter deformation n'1 , n'2 , n'3 , n'4 are marked as Four paired circles (total 8 circles) near apex of this pyramid, drawn just below point O.

With increasing number of protons and increasing atomic number (Z), and the increasing nuclei in atom with increasing atomic weight (A) it gradually increase the size of nucleus, and also forms 1st, 2nd, 3rd and 4th quantum levels in both vortexes in this field.

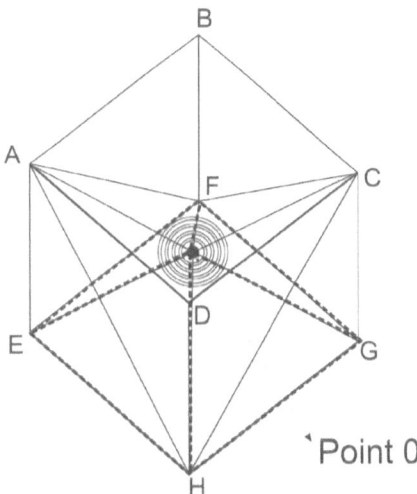

Figure 56: Deformation of cube of one unit space matrix in second dimension starts from point A, while counter deformation starts from point C, but finally both deforms square EFGH. Thus a pyramid shaped area with its peak at point O and its base on EFGH will be over laped by two deformations. Near apex of pyramid thus formed below point O four quantum levels of deformation and conter deformation are marked as 8 circles, where atoms of all known 118 elements condenses.

Thus increasing deformation with inreasing load (protons and nucleons) at four quantum levels will form four shells from each vortex.

1st quantum level **shell** will have one sub shell named as *s; at* 2nd quantum level, it will have two sub shells (*s* and *p*); at 3 rd quantum level three sub shells *(s, p, d); and at* 4th quantum level it will have 4 subshells (*s, p, d, f*); and similarly four shells will contributed from counter-deformation vortex will have sub shells in the same pattern at 1st, 2nd, 3rd and 4th quantum levels.

These 8 quantum levels will form near point O, at centre of 4 paired circles drawn in Figure. The **nucleus of this increasing deformation will enlarge (sink)** along a plumb line drawn from Point O, to 'the meeting point of two imagenary lines AG and HF'.

Every orbit formed can have two opposite spin electrons; Spin Up is clockwise, while Spin down is counter clock wise spin of electron.

In First shell their is one first sub shell s their is 1 orbit which can have 2 electrons (1+1);This is K atomic energy level.

In second shell their are two s, p sub shells will have 1 and 3 orbits; s with one orbit harbouring two (1+1) electrons and p sub shell with 3 orbits has six electrons (3+3); thus total 8 electrons (2+6) are in second shell; These shells are at L atomic energy level.

In third shell their are three s, p, d sub shells will have 1, 3 and 5 orbits; where s has two (1+1) electrons and p has six electrons (3+3), while d subshell has 10 (5+5) electrons totalling 18 electrons (2+6 +10) in these shell; These shells are at M atomic energy level.

Fourth shell will have four sub shells s, p, d, f with 1, 3, 5 and 7 orbits; where s has two (1+1) electrons, p has six electrons (3+3), d subshell has 10 (5+5) electrons and sub shell f will have 7 orbits with 14 (7+7) electrons, and will total 32 electrons (2+6 +10 +14); These shells are at N atomic energy level.

With **each addition of Proton and** alongwith increasing Z number, **equal number of electrons will fill the empty orbits formed in**

both deformation and counter deformation, to form all 118 known elements grouped in periodic table.

The filling of electrons with +spin and −spin in each orbit, obey Pauli's principal, Hund's rule and Aufbau principal.

Seven periods of periodic table, incorporat all 118 elements which are formed in 8 shells at these 8 quantum levels, where first period of periodic table have first two elements which are formed at n1 and n'1 two atomic quantum levels.[24]

With increase of one proton (or nucleon) the deformation of second dimension will shift at its matching quantum level. But atomic quantum levels forms only at quantum levels, where circumference of Electron orbit is equal to the integral numbers of its wave length,

$$J = mvr = nh/2\pi$$

Radius of permitted orbits are proportional to square of quantum number (n), in integrals of 1, 2, 3 and then $r = n^2$ (and r can be 1, 4, 9 and 16 times etc)

Bhor's orbits is at 0.53 A^0; 4 x 0.53 A^0; 9 x 0.53 A^0 and 16 x 0.53 A^0

(where A^0 =10 x 10^{-10} Cm)

1/2 Spin of particle (of nucleons) means that after two full rotations it will return to its origional orientation, while 2 spin of graviton means it will require only half rotation to regain its orientation (like card of

---

24 [At first two quantum levels, each with one '*s*' sub shell it will form one hydrogen and one helium atom. With increased number of Proton from one to two, the pairing with opposite spin Electrons in both s orbits (of n1 , n'1 quantum level) is not practically **possible** and as such twin elements with mirror image configration are not possible. The filling of second electron while formation of helium element it will either add, to form ParaHelium (second electron is paired electron in *s* orbit of n1; of antiparallel spin) or OrthoHelium (second electron is at *s* orbit at other n'1 quantum level in other vortex; of parallel spin)]

a queen or king in a pack of playing cards). Addition of electrons with opposite spin means +Spin and −Spin electron, with clockwise and anticlockwise direction of the rotations.

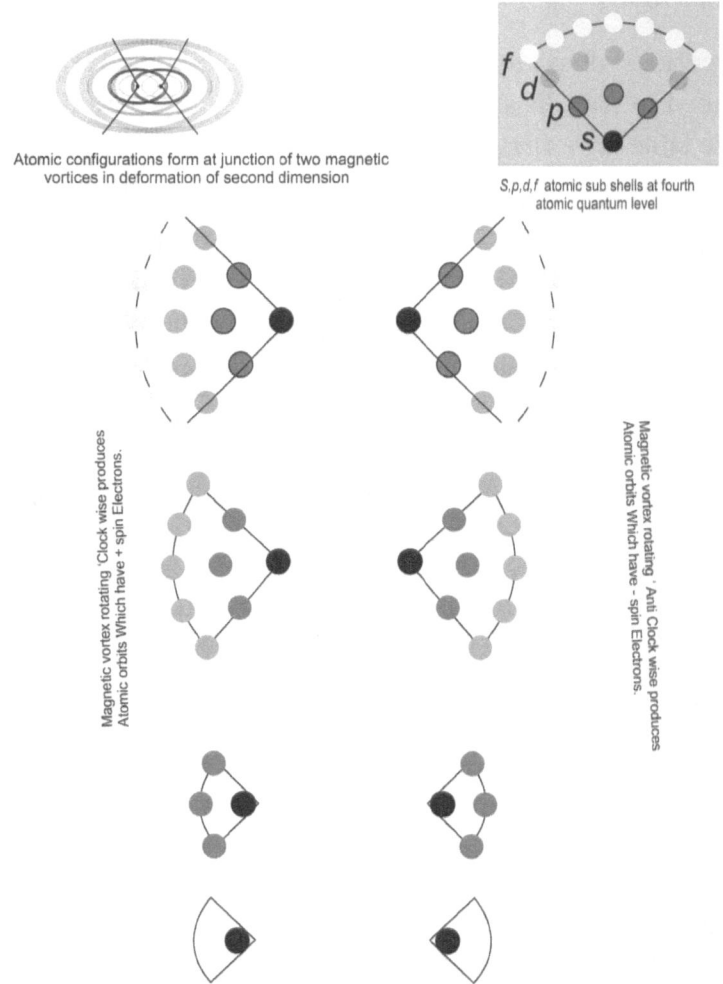

Atomic configurations form at junction of two magnetic vortices in deformation of second dimension

S,p,d,f atomic sub shells at fourth atomic quantum level

Magnetic vortex rotating 'Clock wise produces Atomic orbits Which have + spin Electrons.

Magnetic vortex rotating ' Anti Clock wise produces Atomic orbits Which have - spin Electrons.

Two magnetic Vortexes rotating in 'clock wise' and 'anti clock wise' rotations form Atomic orbits with + Spin and - spin Electrons which totals 2,8,18,32 at four 'atomic quantum levels'.

Figure 57: Orbits at four quantum level with one by one increasing s, p, d, f subshells.

In geometry of each shell in second dimensional deformation, constituent **kinetons** of one rotating kinetic coloumn of shell (with its sub shell and orbits) occupy only **1/3 area of shell, while in remaining 2/3 area of this disc of shell, this kinetic coloumn rotates, to complete circles of all orbitums and the moving kinetons are seen as electron cloud.**

But with condensed universe in era of nucleosynthesis, three kinetic coloumns formed by quarks could have been accomodated in same discs (shells), which formed nucleons. **The orientation of these kinetic coloumns in one out of x, y and z planes will ditinguish them in three classes, indicating three colours of quarks.**

In all orbitum which togather form **any one orbit, few orbitums may come from one rotating vortex and remaining from other vortex. It may be 1/3 and 2/3 division of total orbitums forming one orbit** indicating distribution – 1/3 eV and +2/3 eV charge of quarks. **This is only way to form 'non integral number of charges' (1/3 or 2/3), from 1eV charge produced by all orbitums on one electron.**

Six flavours of six type of quarks only indicate six quantum levels (from two vortexes) at which they condenses according to their respective individual different masses. And when they jump to lower quantum level to change their type (flavour) they shed off their their extra mass energy, like jump of electron in atomic orbits to form photon.

The **particle and anti particle nature is indication of direction of revolution** of particle in its orbit.

All bosons; Photons, 8 type of gluons, W+, W, W – , W (+ –) and Z particles **are basically kinetic coloumns** made up of kinetons to support their different loads at different quantum levels of kinetic coloumns, which form in second dimension. While graviton (boson) is formed at apex of one such fully formed graviton coloumn.

Magnetic fields; The rotating kinetic coloumns of electron (and charged particles) , will form a cloud of **rotating kinetons** in total area of all shells as magnetic field of electron. The rotating orbitals of this kinetic coloumn are magnetic field lines.

Charge; Rotation of the mass energy of Electron and Proton at its core are its **rotating free treos, which decides the charge strength by its number** while its + ve charge or − ve charge is decided by its **direction of rotation.**

The role played by mass in gravitational fields is replaced by charge in second dimension. In general **smaller the mass** of any charged particle, **longer is its length of spread** on RC wave length and proportionately **greater is its charge.** (Body surface of a baby is more with its less body mass).

(b) Atomic Energy and quantum mechanics

HYDROGEN ATOM

1. *Smallest Atom hydrogen is of 1.660539066 × 10^{-27} Kg*

    = **2.62421482 ×10 ^{67} treos, or 1.4143734 ×10^{24} quanta mass energy.**

2. **1.31181654 ×10^{19} bound treos is RC wave length of hydrogen atom.**

    (Then 1.07817694 × 10^5 quanta will be exerted load per apex bound treo in RC wave length)

3. *The energy of hydrogen atom is − 13.6 eV.*

    (as, 1 eV is equal to 1.60218 × 10^{-19} J)

4. *When converted in joule is − 2.178 × 10^{-18} J ;*

    (This value − 2.178 ×10^{-18} J ; is Rydberg constant for hydrogen atom RH)

5. One electron in Bhor orbit of atom is at 'first energy level K' at **Bhor radius (5.29 177210903 ×10⁴ fm is calculated as 32.72252276 × 10²³ bound treo distance; i.e. distance of Bhor orbit from hydrogen nucleus).**

Total energy of hydrogen atom = sum of its Potential energy & Kinetic energy. Where, Potential energy is Electrostatic centripetal force of attraction while its kinetic energy is due to motion of its electron exerting centrifugal force.

*By keeping a balance of these two forces one electron can hover with resultant – 13.6 eV energy in Bhor orbit. (**with ground state of hydrogen atom**) It is calculated as E= e²/2a where e is charge and a is fine structure constant.*

Minus sign from this energy of hydrogen atom is hereby removed. (The theories are some times twisted by all known theorists to get desired results. Now the time has come to give a small twist in our belives, **to see the truth**, We will remove the arbitararly introduced negative sign from energy of hydrogen atom keeping all values and calculation same with same formulas; (*as **arbitrarly introduced negative sign** of this energy, is introduced only to indicate that this amount of energy is equal to the positive ionization energy required to expel the single electron from hydrogen atom*) **hence it is removed in this model.**

(A) This 13.6 eV energy is equal to diluted mass preesure of one proton/hydrogen atom at a distance of Bhor orbit.

(a) Now as 1 eV charge is of one unit electron on 1.439491604×10⁶⁴ free treos contained in one unit electron, then 13.6 eV energy of hydrogen atom will have charge on **19.5770858×10⁶⁴ free treos.**

(b) It is observed and documented here for the first time, that this hydrogen atom energy of 13.6 eV or **charge on 19.5770858×10⁶⁴ free treos.**

when **divided by fine structure constant (1/ 137)** = is charge on 2.6 (8206075) × 10^{67} free treos, which is ≈ mass energy of one proton or of one hydrogen atom.

It is ≈ **mass of hydrogen atom which is 2.62421482×10^{67} treos.** ≈ **mass energy of one proton which is 2.64416818×10^{67} treos).**[25]

Thus not only one + eV charge of proton is netralized by one − eV charge of electron; **but at bhor orbit the exerted mass pressure of one proton /hydrogen atom = 19.5770858×10^{64} free treos is neutralized by 19.5770858×10^{64} kinetons,** present with one **electron in Bhor orbit.** (as 13.6 eV energy of hydrogen atom at this K energy level)

Thus this 13.6 eV, the energy of electron, at Bhor orbit (i.e. at K energy level) is to neutralize; the mass pressure of one proton in nucleus of hydrogen atom, at this distance of Bhor radius.

The role of mass in gravitational forces is replaced by charge in second dimension to describe atomic forces, and thus it indicates some ACTION − REACTION MECHANISM i.e. quantum gravitation in these atomic fields.

{(!) We know that each free treo has **minimum possible mass energy = 0.632570165×10^{-94} Kg weight per free treo.** (as, 1/ 1.5808523×10^{94} Free Treos are in one Kg); page 48

(!!) and each free treo also have **minimum possible charge of 1.11261604×10^{-83} coloumb per free treo** (=1/ 0.898782656 × 10^{83} free treos per coloumb); page 156

(!!!) *and this is in accordance with the known value of Charge mass ratio = 1.7588 × 10^{11} coulomb/Kg;* page 156}

---

25 we can note that *Proton mass is more then hydrogen atom mass* (proton mass + electron mass) *as this energy difference is fusion energy used to form one hydrogen atom.*

(B) **13.6 eV Hydrogen atom energy is equal to value of Rydberg constant RH 2.178×10⁻¹⁸ J** *(minus sign removed as discussed before)*

The 13.6 eV energy of Hydrozen atom is equal to 2.178×10⁻¹⁸ J, It is Rydberg constant for hydrogen atom RH. (1 eV is equal to 1.60218×10⁻¹⁹ J)

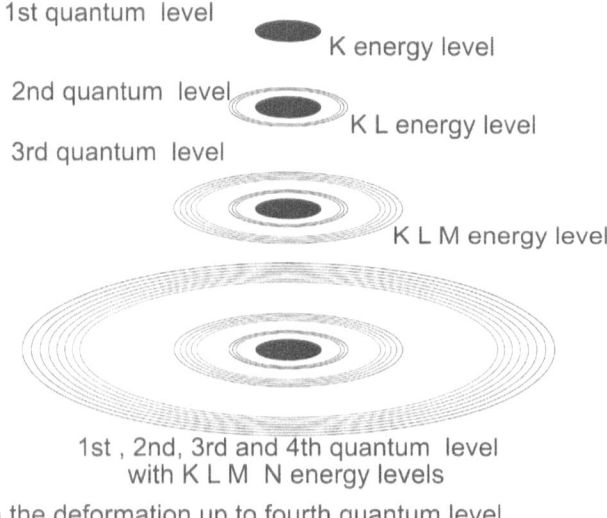

1st quantum level — K energy level

2nd quantum level — K L energy level

3rd quantum level — K L M energy level

1st , 2nd, 3rd and 4th quantum level
with K L M N energy levels

In the deformation up to fourth quantum level,
previous deformations at 1st , 2nd, and 3rd quantum
levels also combines tegether to form atomic structure.

Figur 58: Orbits at four quantum level with increasing number of *s, p, d, f* subshells, representing K, L, M, N atomic energy levels.

With increase of one proton at nucleus this energy of 13.6 eV is achieved by all orbits present in one newly added sub shell formed, according to known formula En = 2.178×10⁻¹⁸ J Z²/n².

[By applying Z (Atomic number) and n (primary quantum number) in this formula.]

After removal of −ve sign now above calculated energies match exectly with the proposed geometrical pattern of deformation in second dimension geometry. With the use of positive values we can demostrate action-reaction mechanism in atomic fields, in accordance with quantum gravitation.

The increment and distribution of this Hydrogen atom energy 2.178×10^{-18} J as rydberg constant can calculate base energy for all other atoms at first atomic quantum level n1 as energy K by known formula is E1 = 2.178×10^{-18} J Z^2, and for its distribution at other quantum levels n2, with L energy, n3, with M energy and n4 with N energy levels it is calculated for each one orbit by known formula, En = 2.178×10^{-18} J Z^2/n^2, or by another simple known formula E= E0 $/n^2$.

THIS ENERGY DISTRIBUTION IN A SHELL (in one kinetic coloumn as per coloumn geometry) was discussed on page 90, as distribution of load in one kinetic coloumn and ENERGY DISTRIBUTION IN AN ATOM on page 210 (distribution of mass pressure of body in gravitational field) and its reaction (indicated by speed of electron). This justify our removal of −ve sign from hydrogen atom energy value and from all its formulas.

(1) In an atom with one proton at **n1 quantum level, at 'K' energy level**, the energy present in its 1 orbit, of s sub shell = E1 = *2.178×10^{-18} J ; this energy is RH Rydberg constant for hydrogen atom used as constant in known formula 2.178×10^{-18} J Z^2.*

Speed of electron is about 2200 km /sec

(The formula is valid for hydrogen like atoms, but calculates the energy to be acquired by an electron to fill an empty orbit in other elements also).

(2) In other **atom with two protons (Z=2), at n2 quantum level** with K and L energy levels, energy at K energy level will be E2 = $2^2 \times 2.178 \times 10^{-18}$ J = and will divide at (second) **L energy level** as per known formula $E= E0 /n^2$ in each of 3 orbits, of p sub shell = E 2 /4 = *2.178×10^{-18} J (will be same).*

But speed of electron becomes 1100 km /sec

(3) In the **atom with three protons** (Z=3), **at n3 quantum level** with K and L and M energy levels, energy at K level will be E3 = 3^2 × **2.178** **×10^{-18} J** and will divide, at (third) **M energy level**, as per known formula $E= E0 /n^2$ in each of 5 orbits, of d sub shell = E3 /9 = **2.178** × 10^{-18} J *(will be same)*.

But Speed of electron is about 2200/3 = 733.3 km /sec

(4) Finally in **atom with four protons (Z=4)**, **at n4 quantum level** with K, L, M and N energy level, energy at K levels will be E4 = 4^2 × **2.178** × 10^{-18} J = and will divide, at N (fourth) energy level, as per known formula $E= E0 /n^2$ in each of its 7 orbits, of f sub shell = E4 /16 = **2.178 × 10^{-18} J(will be same)**.

And speed of electron is about 2200/4 = 550 km /sec

This energy need not be confused and compared with different ionization energies of atoms and assosiated with particular electron.

The ionization energies are quite variable as it changes its value, as the particular electron is effected by "shilding effect of electrons in inner orbits"; 'repulsion of other electrons in same shell' ; 'half filled or full filled shells', depending on this particular electron position in shell.

(a) *The place of Electron orbits is decided by a delicate balance of electrostatic force of attraction (centripital force) and (centrifugal force) by electron motion in its orbit.*

(b) *Thus each orbit should have of 13.6 eV or 2.178 × 10^{18}J as minimum* **energy of electron, required to match the diluted mass pressure of proton at this level of 13.6 eV or 2.178 × 10 ^{18}J.**

(c) *But one new electron will be captured only when 1 eV + charge is increased at nucleus by addition of one more proton.*

Similarly as a further proof, when this energy is counted at K L M N energy levels in s, p, d, f orbits, in one mol of hydrogen atoms it calculates $1.312 \times 10^3 K J$ per mole, as base energy at K energy level (= 2.17×10^{-18} J multiplied by Avogadro number 6.022140×10^{23} hydrogen atoms: minus sign removed); *and then it is divided at four quantum levels from E to E/4, E/9 and E/16.* These observed calculated (known) values can also be *explained by treo model.*

K energy level = $1.312 \times 10^3 k J mole - 1 = E$

L energy level = $3.2 \times 10^2 k J mole - 1 = E/4$

M energy level = $1.46 \times 10^2 k J mole - 1 = E/9$

N energy level = $82 k J mole - 1 = E/16$

(C) Catapult of nature

13.6 eV ground energy of hydogen atom *also drives the value of Rω Rydberg constant in wave numbers '1.09677.57 cm^{-1}*

Rω Rydberg constant in wave numbers 1.09677.57 cm^{-1} **is used as a constant** in formula [Rω $(1/(n1)^2 - 1/(n2)^2$ cm-1] to calculate the wave number (1/wave length) of all photons, responsible for producing absorbtion and emission spectra in all series and by any element.

Lyman, balmer, paschen, brackett and pfund series lines in spectroscopy formed by emission or absorbtion of energy from atom, which is, received or supplied by a photon of fixed wave number. [i.e. wave number (1/ wave length) of photon emitted = c h /change in electron energy.] Atom of any element absorbs the photon at particular wave length which mark as black line (dip) in absorbtion spectra and releases the photon of **same wave length and energy** which mark as white lines (spike) in emission spectra.

By jumping of electron from orbit at higher energy level to orbit at lower energy level in an atom of any element the energy difference is relesed as photon Quanta. **Bigger the jump of electron from higher to lower energy** *orbit (or vice versa) the high energy transiction is done by trasfers of higher energy mass photons, while smaller jumps or lower energy transiction is done by lesser energy mass photons. The more you pull your catapult the bigger stone you can throw.* The balmor series produces photons in visible range of wave lengths 400nm to 700nm (hydogen atom emits photons in visible range at 484 nm wave length).

These properties provides any element its unique identity and color. This fixed energy released or absorbed, also assign chemical properties to this element, and thus constitution of outermost electron orbit is responsible for all physical and chemical properties of any element.

Thus it can be noted, that an element and molecule react to its surroundings, in response to stimulus of photon, by changing the frequency of recoiling photon (Raman effect or compton scattering etc.) or expresses itself in a calculated manner with release of photon of its unique colour.

All elements absorb and release energy, as photons at its unique wave length, which can be compared with all living being of different species, (which reacts to its surroundings, eat according to prefences of species and releases its unique wastes). Are not all this indicates the generation of some prototype of cociousness at atomic level.

(D) Formation of an atom

(1) **UNIT ELECTRON** (conventional values and values as per treo model)

1. *Mass of unit electron = $9.1 \times 10^{-31} Kg$* = **$1.439491604 \times 10^{64}$ free treos mass energy.**

2. *This mass energy of unit electron spreads on its RC wave length of $(0.38600391 \times 10^{-12} M)$* = **$23.88951 \times 10^{21}$ bound treos length.**

3. *Compton wave length $(2.4263102 \times 10^{-12} M)$* =**$1.5016263 \times 10^{23}$ bound treos length** *(R C wave length $\times 2\pi$)*

4. *The charge energy of unit electron, **1 eV or 1.6 × 10⁻¹⁹ J**.*

5. *In hydrogen atom one electron is **in Bhor orbit**.*

    (Bhor radius [26] = RC wave length of Electron × 137)

6. ***Hydrogen atom moves in bhor orbit at 1/137 of speed of light*** *and that is also the speed of electrons on wire and on neurons i.e. about 2200 Km per second, while its charge moves near the speed of light without actual movement of electron at that speed.*

(explaination – imagine a queue of persons in front of a restaurant and persons one by one reaching the counter, now although the last person will reach the counter much later, but if this last person pushes the person in front of him, this push will reach as wave quickly to front/to first person in line).

7. ***Single electron around nucleus, of any element*** *with its atomic number Z, in its matching orbit moves at **Z/137** times the speed of light.*

8. Electron can not fall in nucleus, being separated from it by 10^4 quantum levels (10000 times away) as electron is present at 10^{22} quantum level while nucleons are present at 10^{27} quantum level.

(2) PROTON

Conventional teaching describes, Protons are made up of fast moving quarks bound togather by gluons (as they are bosons, gluons are kinetic columns formed by kinetons; in atom they works as cementing matrial).

---

26 *(a) If you divide RC wave length of unit electron by fine structure constant 1/137 you will get the **Bhor radius** of 5.29 177210903 x 10⁴ fm or **32.72252276 × 10²³ bound treos radius** (b) If you multiply RC wave length of unit electron by fine structure constant 1/137 you will get **classical radius of electron** 2.82 fm or **1.7437599 × 10²⁰ bound treos radius**.*

The proton energy of 938 MeV is mostly its **quantum chromodynamics binding energy** *and it's mass is 1836 times of unit electron mass.*

(Mass energy of electron; **0.511 MeV × 1836 = 938.196 MeV;** *Mass energy of Proton)*

1. *Mass of Proton* $= 1.6726219 \times 10^{-27}$ *Kg*

   $= \mathbf{2.64416818 \times 10^{67}}$ **free treo** $= \mathbf{1.42\ 512458 \times 10^{24}}$ **quanta.**

2. *Proton have positive charge* $+ 1eV = 1.602176634 \times 10^{-19}$ *coloumb,* and it is exactly equal to charge of electron in magnitude, but has positive charge.

3. **Proton charge radius** *of 0.875 fm (may be 0.842 Fm or even 0.833 Fm) when converted in bound treos* $(0.875$ fm $\times 6.188957 \times 10^{19}$ bound treos per fm $= \mathbf{5.415337 \times 10^{19}}$ **bound treo radius**$)$; It is calculated by electron scattering experment (by which we determine the area of charge distribution and its pattern i.e, + charge or − charge on the sphere), **and it is found that Proton surface area is mostly positive.**

4. HEIGHT OF PARTICLE Length of spread of (vertically placed) Proton packet = **RC wave length** $= \mathbf{1.30547046 \times 10^{19}}$ **bound treo length.**

5. BREADTH OF PARTICLE The Proton packet exert a very heavy load of $\mathbf{1.42\ 512458 \times 10^{24}}$ quanta at each apex bound treo, which is supported by one shell (kinetic coloumn) of $\mathbf{1.42\ 512458 \times 10^{24}}$ bound treo radius or bound treo layers (frequency of Proton).

With positive charge an Proton it can capture a electron in its first sub shell at Bhor orbit, and thus it gets converted in to hydrogen atom. This proton capturing a electron to form an stable configration of atom; (can be seen as lady capturing a bread earning husband to live in a stable home). Electron by its constant motion constantly accumulates desired energy in atom which will also soon decipates in its surrounding or will be released by emission of a photon from atom; This electron capture will occure at $3.272252276 \times 10^{24}$ bound treo distance in Bhor orbit, placed at more then double the size of proton disc.

(3) NEUTRON

a. *Mass of Neutron* = $1.67492749 \times 10^{-27}$ *Kg* = **2.64781297× 10^{67} free treos**

   In a layman language, we can count Neutron as one proton attached with one electron, with its added charge of − 1 eV , which neutralises positive charge of proton of +1 eV (1eV = $1.602176634 \times 10^{-19}$ coloumb.

b. *The charge radius of Neutron is 0. 8 fm and charge distributions pattern (by electron scattering experiments) of neutron is **positive core at centre and mostly negative in periphery.***

c. *Neutron is unstable and by beta decay it converts in proton by releasing a **beta particle** (fast moving electron; along with its supporting W boson), which in turn gives 1 eV positive charge to proton and thus make this Proton capable to catch a electron of 1eV negative charge.*

(4) QUARKS

The **quarks has never been found as independent elementary particle, and can not be seperated from each other, when as a constituents they form composite particle and as such no point charge can be created;** thus in this model, Quarks (6 Quarks and 6 antiquarks as its flavours) are suggested as **architectural patterns of composite particles of nucleons,** rather then one seperate class of independent elementary particles.

1. Due to different mass energies of 6 Quarks, formed in three generations (First generation, **Up – Down:** second generation,

Charm –Strange; and in third generation, **Top –Bottom**) they condense at increasingly higher six quantum levels.

2. In Quarks (as also elsewhere), the **anticlockwise rotation of free treos spread (one free treo on one kineton) on their kinetic coloumns** along its axis gets **their – ve charge which is inwards (centripetal flow) and Quarks rotating clockwise gets +ve charge with outwards (centrifugal flow).**

3. The quarks of all **three generations** are grouped according to their contribution of charge 1/3 eV – ve charge (in Down, strange & bottom) and 2/3 eV + charge (in Up, Charm & top). Their **six flavors only describe their formation at 6 different quantum levels** in kinetic coloumn of second dimension. Four quarks of second and third generation quickly degenerates to be conveted in Up or Down quark of first generation which formed all matter of universe. **Only Up quark with 2/3 eV positive charge and Down quarks with 1/3 eV negative charge form all nucleons and mesons.**

4. Their **particle and anti particle status** is determined by their counter clockwise and clockwise **revolution** of quarks in its individual orbits.

5. Quarks with Spin ½ will (**rotate twice while forming its one matter wave**) to regain it origional orientation. – 1/2 is down spin with anticlockwise rotation and +1/2 is up spin with clock wise rotation.

6. In matter wave, the particle spreaded on its RC wave length rotates once (by one by one vibrations of all shells from top to bottom of vertically spread particle in its RC wave length) and **then** particle shifts to next one bound treo in its orbit. The particle rotates once more (by bottom to top vibrations) and then again shifts to next bound treos in orbit, and thus it form one matter wave on 2 RC wave length. **After two rotations the**

particle with spin 1/2 regains its origional orientation. This shifting to next bound treo after its one rotation of particle, it decides speed of its orbital motion.Page 237.

7. Quarks are produced in this pyramid Shaped area formed by overlapping of two rotating vortexes A and B, with clockwise and anticlocwise direction of rotation of their sheets and they contribute 1/3 number of orbitums of minus charge and 2/3 number of orbitums of positive charge.

8. Three flavours of quarks forms in each vortex at three resonant quantum levels, to form 6 quarks of three generations. The (down, strange, bottom) quarks are form in A vortex with negative charge while (UP, charm, top) are from B vortex with their positive charge.

9. **One Down quark from A vortex will contribute for kinetic coloumns of 1/3 √S number of orbitums, and two UP quarks will contribute 2/3√S + 2/3√S number of orbitums from vortex B, while forming proton mass with resultant net + 1 eV charge.**

10. **One Up quark from B vortex will contrabute for kinetic coloumns of 2/3 √S number of orbitums and two Down quarks with 1/3 √S + 1/3 √S number of orbitums will contribute from vortex A to form neutral Neutron mass.**

11. Quarks only contributes for roughly 9 percent of nucleon masses. While Up quark is of 5 Mev/ c^2 and down of 10 Mev/ c^2, while Proton of 938 MeV/ c^2 are formed by **27 C mue (composite mass of unit electron)** units each of 35.12 MeV/ c^2 ; See page 47, (for comparison Electron have 0.511 MeV/ c^2) .

12. **With its fluctuating kinetic energy 0 , – (minus) or + (plus) of each orbitums (Fig. 11) they get three colours, reflected by the orientation of the orbitums in 3 spatical directions**

i.e. x, y, z plains, while forming one Proton or Neutron shells along their RC wave length.

Thus these architectural patterns of energy condensation at one quantum level in vortex A and one quantum level in vortex B, of Down and Up quark in turn regulates the shape, mass and charge distribution of all composite particles to which they form.

(5) WEAK FORCES and W , W +, W − , W+ − AND Z PARTICLES

In this hyper dense area of Atom, W +,W − (of mass 80.379 GeV/c²) and slightly bigger Z particles of intermediate vector boson (91.187 GeV/c²) are **mega size supporting kinetic coloumns,** which are formed with or without charges in pyramid shaped area of deformation and counter deformation in second dimension, to support densely packed nucleons and sub nuclear units (alpha particles) and also due to their dualistic nature, they act as vehicles for their transportation as waves.

Alpha particle is sub nuclear unit of four nucleons (a Helium nucleus) and it is known to have produced 99 % Helium on Earth, while **beta particle** is energetic electron moving at very fast relativistic speed; which is emitted in beta decay of neutron. [Beta decay is conversion of Neutron to Proton with release of one beta particle carried away by one W boson, as it is also released].

While **gamma rays** the third type of radio active emission occur from atom in fission, fusion, alpha decay or gamma decay.

Here 10^{16} bound treos is RCwave length of these bosons of weak forces, matches with **range of action** of these bosons of weak forces. They could execute these transportations by **converting alpha and beta particles in to a wave** which crosses the atomic barrier by **quantum tunneling,** i.e. by spreading all treos forming this alpha particle, on equal number of transporting kinetons in W or in Z bosons (it is explained in detail as wave particle duality on page 71).

Like photons intermediate vector bosons of W and Z particles have spin 1 (as they also regains their orientation with its one rotation on one wave) and are **not very different from photons except they condense at 10^{27} Quantum** level in second dimensional deformation.

Having charge they can not accumulate in increasing numbers as beam of photon, thus the weak forces remains weak while EM forces by piling of photons can become very strong.

(E) Formation Of Nucleons

We saw that mass energy accumulated by S free treos (one quanta mass energy) at each of √S quantum levels of first dimension, and formed all photons of EM spectra. While in second dimension, √S quanta mass energy or √S × S free treos (but not unit electron itself) increases at each of quantum level to form all energetic electrons and elementary particles. But in the **union of full unit electron packets**, to form all known composite particles and nucleons, the question arises how all negatively charged unit electrons could be packed togather in a packet ?

The answer lies in the fact each unit electron (one brick) is associated with **68.5 times of its mass** (as its cementing material) which are bound treos from space matrix and both togather form one **Cmue of 35.012 MeV.** Integral multiples of such **Cmue** (composite mass of unit electron) form all known elementray particles and nucleons (page 47, table I).

It indicates that all matter in universe is made up of integral number of these Mass units (i.e. unit electrons wrapped with space matrix). Such 27 Cmue units will form one proton; **27** × 35.012 ≈ 938.27 MeV as Proton mass energy.

Free Protons are stable particle with its half life 10^{32} years while free Neutron or bound Neutron is a unstable particle and transmutes by beta decay in proton in 881 seconds by emitting a fast Electron i.e. beta particle.

With the formation of all elements, the coloumn geometry prevails as it forms four pairs of shells, (1) s, and s' (2) s, p and s', p' (3) s, p, d and s', p', d' (4) s, p, d, f and s', p', d' ,f' sub shells in which the number of orbitals are one in s, three in p, five in d and seven in f and each orbit will be filled with one pair of electrons, one + spin electron and one – ve spin electron.

With the addition of one proton at nucleus one electron fills up according to set rules of Pauli's, Aufban's and Hund's.

According to second dimensional deformation with the addition of one proton (or neutron) the mass energy of nucleus will increase and it will now shift at a higher quantum level. All protons (each proton with its mass energy spread on its RC wave length) in two rotating vortexs as twisted **double strand** can be accomodated in conventional size of nucleus.

These observation confirm –

(1) With each rise of mass energy e.g addition of one proton and one electron in nucleus to form next element **the nucleus will shift to higher quantum level as Sinking nucleus (in bigger pit of space matrix).**

(2) In this geometry the double strands formed in nucleus will not only include protons but actually **they will make combination of proton, neutron, mesons, bosons of weak forces and gluons** to form two identical revolving vortexes in atomic double helix (an precurser to structure of DNA).

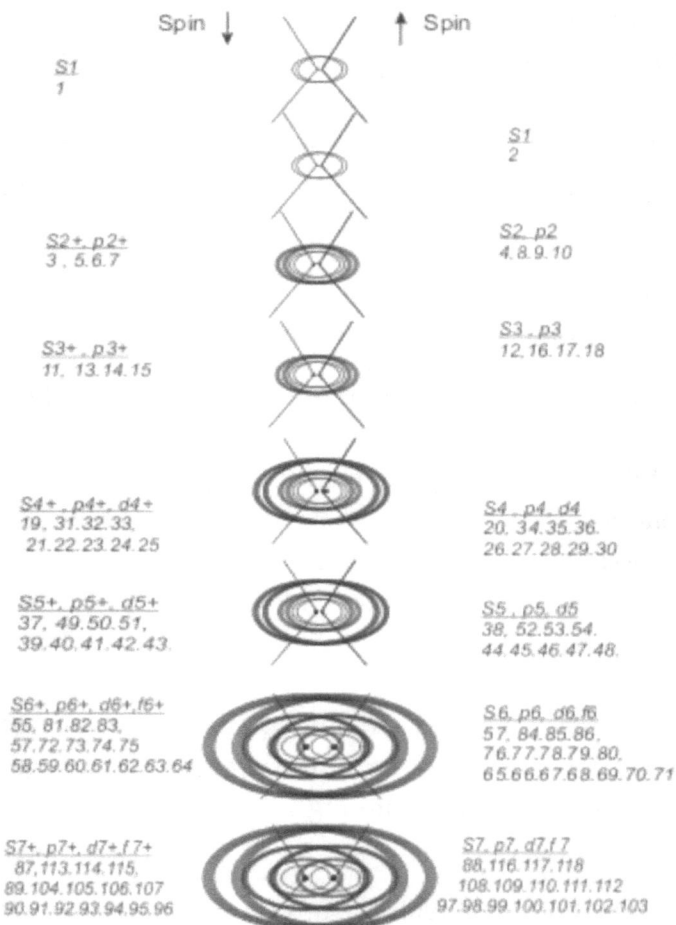

Spin ↓   ↑ Spin

S1
1

S1
2

S2+. p2+
3 . 5. 6. 7

S2. p2
4. 8. 9. 10

S3+. p3+
11. 13. 14. 15

S3 . p3
12. 16. 17. 18

S4+ . p4+. d4+
19. 31. 32. 33.
21. 22. 23. 24. 25

S4. p4. d4
20. 34. 35. 36.
26. 27. 28. 29. 30

S5+. p5+. d5+
37. 49. 50. 51.
39. 40. 41. 42. 43.

S5 . p5. d5
38. 52. 53. 54.
44. 45. 46. 47. 48.

S6+. p6+. d6+. f6+
55. 81. 82. 83.
57. 72. 73. 74. 75
58. 59. 60. 61. 62. 63. 64

S6. p6. d6. f6
57. 84. 85. 86.
76. 77. 78. 79. 80.
65. 66. 67. 68. 69. 70. 71

S7+. p7+. d7+. f7+
87. 113. 114. 115.
89. 104. 105. 106. 107
90. 91. 92. 93. 94. 95. 96

S7. p7. d7.f7
88. 116. 117. 118
108. 109. 110. 111. 112
97. 98. 99. 100. 101. 102. 103

Coloumn geometry (1,3,5,7..) in coloumn and counter coloumn, regulates filling of electron orbits in s,p,d,f sub shells in atomic structure of all elements

Figure 59: At four quantum level with increasing number of s, p, d, f subshells, all orbits are filled with electrons to form all 118 elements of periodic table.

(F) Quantum Gravitation In Relation To Atomic Structures

As we have observed, that any 'm' unit masses body exert a load of m^2 unit masses at its gravitational centre (Action) which is neutralzed around its centre by m^2 gravitons (Reaction) which thus support this mass. This load with distance; dilutes all around in all directions and in all planes as diluted mass pressure at quantum levels of all these (of each graviton) graviton coloumns (which togather join to form gravitational field of this body).

At any one layer of this gravitational field the mass pressure (of central load) is 2 MG; and at any n bound treos distance (radius of coloumn) this 2 MG load spreads uniformally on 2n-1 **apex bound treos** on this one layer on one wave and is neutralized (Reaction) at each apex bound treos by formation of kinetic coloumns.

Simlarly in atomic quantum gravitation Proton made up of 27 C mue or mass units (or $2.62421482 \times 10^{67}$ treos in hydrogen atom) will exerts a load of $(27 \text{ mass units})^2$ at centre of nucleus, and $(2 \times 27 \text{ mass units}) = 2$ proton mass (or 2MG), will be its mass pressure on each layer of atomic coloumn. At a distance n (radius of atomic coloumn, **where value of n is = '137 RC wave length of Electron')** at bhor orbit, It will exert this 2 proton mass $(2 \times 27 \text{ mass units})$ pressure on one layer of $(2n-1)$ apex bound treos on which one wave is formed.

There fore this **2 proton mass** is pressure on **2 x 137 RC wave lengths of electron – 1.**

Thus mass, (of proton mass $/137 = 13.6$ eV hydrogen atom energy, will be **on one RC wave length of one electron in bhor orbit =19.5770858 $\times 10^{64}$ treos,** (as calculated above on page 171).

Thus derived, that the mass pressure of hydrogen atom (or of one Proton) exerted at Bhor orbit, is neutralised by energy of one electron in Bhor orbit, in one type of action –reaction mechanism.

The important mile stones of QUANTUM GRAVITATION

We note here, that the increasing **load of free treos at one apex bound** (or at gravitational centre of body) **is suppoted by equal number of kinetons in the kinetic coloumns (or it may be gravitons)** thus formed; and the number of kinetons in such supporting one kinetic coloumn will be in range from **1 kineton** to S^4 **kinetons** or even more depending upon the load it has to support.

(1) To support one unit Photon; each kinetic coloumn will have **1 kineton (converted from one bound treo).**

(2) To support one unit Electron; each kinetic coloumn will have **S kinetons** (present in **one orbitum**)

(3) To support one Proton; each kinetic coloumn will have $\sqrt{S} \times S$ **kinetons** (present in **one electron at Bhor's orbit**).

(4) To support one unit mass; each kinetic coloumn will have S^2 **kinetons** (present in **one graviton coloumn** placed at one graviton).

(5) To support load of one billion Kg body (S^3 free treos) ; each kinetic coloumn will have **(S^3 kinetons)** (present in one **electron black hole**).

(6) To support exerted load of S^4 free treos, the kinetic coloumn will have **(S^4 kinetons)** present in **gravitational sphere** of **one unit Black hole.**

These kinetic coloumns are working units which produce quantum gravitational fields in all four dimensions of space matrix (space – time).

Thus in case of Proton the diluted mass pressure at Bhor orbit, is neutralised by energy of one electron in Bhor orbit.

We achive our goal of explaining atomic quantum gravitation by proving this relation in all elements; with the help of **existing formulas of energy distribution which are according to geometry of kinetic coloumn in treo model.**

To calculate 'm' or base energy at nucleus of any atom; *formula is E1 = 2.178 × 10^{-18} J Z^2 (where Z is number of Proton in nucleus). This **load increaseses by square (Z^2) times of proton number at centre.** It will now* *be distributed according according to formula En =2.178 ×10 $^{-18}$ J Z^2/n^2 which obey to coloumn geometry for electron orbits.* **Thus each orbit will have 2.178 × 10 $^{-18}$ J energy.**

According to above formula for 2 proton the load at centre will be 2^2 x (*2.178 ×10 $^{-18}$ J*) and its diluted mass pressure will be supported by 4 electrons. (1+3 =4).

Increased 4 times energy at first quantum level (K energy level) will be distributed in 4 orbits.

For 3 protons the load at centre will be 3^2 × (*2.178 ×10 $^{-18}$ J*) and its diluted mass pressure will be supported by 9 electrons (1+3+5 =9).

Increased 9 times energy at first quantum level (K energy level) will be distributed in 9 orbits.

For 4 proton the load at centre will be in square 4^2 (*2.17 × 10 $^{-18}$ J*) and its diluted mass pressure will be supported by 16 electrons (1+3+5+7 =16).

Increased 16 times energy at first quantum level (K energy level) will be distributed in 16 orbits.

With increasing number of protons one by one at necleus the deformation increases to form one by one 8 shells at paired four quantum levels and its energy distribution at L, M and N enery levels is according to coloumn geometry. But empty orbits formed, are filled by **negative charge** electron one by one, only when a **positive charge** proton is added at nucleous.

The number of electrons and their pattern of placement, shape of outer most electron orbit of the atom of any element, which indicates its valiancy and are responsible for all chemical and physical properties of this element; or in other words the **size, shape and structure of space matrix deformed by an atom is responsible** for all chemical and physical properties of this element.

5. Sharing of Kinetic Coloumns of One Body by Other Body

Sharing of kinetic coloumns between atoms, ions or molecules produces **chemical bonding, while electrostatic attraction and repulsion is centripetal and centrifugal forces produced by different direction of rotation of these layers (see page 158). In ionic bonds electrons are transferred and in covalent bonds they are shared and same is in metalic bondings.** While this sharing in intra and inter molecular spaces unites molecules (and thus develops **van der walls force).**

6. Transfer of Full Layers of Kinetic Coloumns

Any energy transfer can only occur in units of integral number of quanta.

This transfer of quanta energy from photon, electrons, any elementary particles, any thermodynamic transfer of heat, transfer of kinetic energy which increases speed and momentum, **any biological process** or any chemical energy transfer **is transfer of integral number of full kinetic layers from all coloumns in wave of donating body to receiving body.**

Or full photon or electron packets are transferred along with its supporting kineton layers; even transfer of gravitons along with their respective graviton coloumns is transfer of gravitational forces when a body comes in gravitational field of other body.

Transfer of each one layer takes one Planck least time; which is one least processing time of universe. **Thus it takes some time to cook the food, to accelerate a vehicle or to complete a chemical reaction or for any biological process to take place.**

In mechanics we are used to the idea that the kinetic energy of a body can be converted totally into work; as all kinetons which are transfered with the transfer of kinetic layers are added to the kinetic coloumns of receiving body, in kinetic energy transfer, to increase the speed or momentum.

But this 100% efficacy is impossible for heat. When Thermodynamic coloumns layers are transferred from higher temperature gas to lower temperature gas, only some unfixed percentage of energy is utelised in work done (the effeciency of engine). **Minimum discrepancy of length of kinetic energy layers donated from host molecule to the size of recepient kinetic coloumn in recipient molecule increases the efficacy of system (engine).** The transfer of layers in coloumn or the temperature difference *decides the effeicacy of engine* $\eta = (T1 - T2/T1)$. Low temperature combustion engine must be effecient engine.

In this process the portion (of one or few quanta) of donated layers from kinetic column of donating body, which does not match with the coloumn geometry (size of layers of receiveing kinetic coloumns) of receiving body, goes to thermodynamic sink (our space matrix); Page 288, Thermodynamics chapter BOOK, INSIDE A WAVE [ref 8].

All photons move as waves only after distributing its total mass energy on their supporting kinetic coloumns (@ one free treo of mass energy on one kineton, in all supporting sub kinetic coloumns). Thus in all kinetic energy transactions some quanta of mass energy as free treos are also exchanged (riding with the transferred layers of kinetic coloumns) which increases mass of fast moving body (one postulate of theory of relativity) or are released as one neutrino, out of Neutrino family.

In Raman Effect (by which we identify molecules) few fixed number of layers from kinetic coloumn with some quanta of mass energy of interacting photon is stolen or donated; and after this interaction the **changed frequency of photon** marks the type of molecule with which it interacted.

While in the process of 'Compton scattering' and in 'inverse Compton scattering', the change in frequency of scattered photon or electron is due to transfer of **few kinetic coloumn layers from all sub kinetic** coloumns of incident photon, which are either stolen from photon or donated by energetic charged electron, during 'electron photon interaction'.

Heat transfer of full layers of kinetic coloumns of photons in infra red frequency; or while transfer of few layers of kinetic coloumn from heated body increases additional layers in kinetic coloumn of any element which thus increase the internal energy of element indicated by rise of its temperature. The heat transfer by conduction or convection of heat are transfer of layers of kinetic coloumns (from heated excited molecules to cool receipient molecules).

The temperature is related to the energy per degree of freedom directly proportional to number of layers in (thermo dynamic) kinetic coloumns of gas molecules, which decide both volume and temperature of gas; (thus it required van der wall correction of gas law) while number of degree of freedom (number of kinetic coloumn in wave) multiplied by mean energy gives the total energy of the system. Degree of freedom can be characterized, given one condition, by the frequency or energy of motion.

Temperature increases in steps and have a scale of 10^{22} units (for convenience we have changed the temperature scale in Celsius degrees) it clearly indicates $\sqrt{S}$ quantum levels are also in thermodynamic coloumn.

The coloumns of EM forces can pile up as photon beams with increased density of field and number of its kineton layers along with increased height of coloumns (as layers of magnetic fields increases with increasing magnetic flux). When photon gas accumulates along with piling of its transverse coloumns it increases proximity and number of kineton layers, which increases strength of EM forces (in square), but without rise of temperature because Photons can compensate the increased pressure by changing frequency of radiation.

These gas molecules while performing its Brownian motion will now strike on the walls more forcefully and frequently which increases the pressure on wall of container. Over lapping kinetic coloumns in reduced area which increases number of kinetic layers per unit area, and in turn it will increase the kinetic energy and height of kinetic coloumns (temperature) of all gas molecules.

Vice versa increase in temperature (increase in height of kinetic coloumns of all gas molecules, by transfer kinetic coloumn layers from thermodynamic coloumns of absorbed photons to gas molecules as heat energy) will also lead to increased pressure of gas (if volume is fixed). This relation ship is governed by Gas laws. Boltzmann's constant gives relationship of temperature of gas with total energy of gas molecules.

7. Spread of Load of Moving Bodies of Multiple Unit Masses, in Time

What is universe? It is pool of energy, in which its kinetic energy imparts, S vibrations per second to all bound treos in space matrix, for total duration of S **seconds, in one life span of this universe.**

It express itself, as S free treos or **one quanta unit mass** of one unit photon, spreads on S bound treos in **one unit space**, and will be supported by S kinetons i.e.**one unit quanta kinetic energy,** by its S vibrations which will occur **in one unit time** of one second.

As 1 free treo is supported by 1 kineton by its 1 vibration for 1/S second. To support one free treo for one second S vibrations (in one second) will be needed.

The observations are, total mass of moving photon of n quanta, will spread on S/ n apex bound treos in its wave length. The load of this mass will be supported by S/n vibrations for S/n vibration time. To support this photon of n quanta for one second (one unit time) it forms n waves (as frequency of photon is n) in one second.

It will consume S/n x n =S vibrations to support n quanta mass energy of photon packet in one unit time of one second. Thus **one wave supports one quanta mass energy** in one second (S vibrations) as **n quanta mass energy is being supported by n waves in one second** i.e. S vibrations.

Similarly any elementary particle packet spread in second dimension on its RC Wave length (supported by one shell at each apex bound treo where its each **quanta mass energy (S free treos) is supported by one orbitum (made up of S kinetons) by its one rotation in one second (S vibrations).**

In both cases the body made up of any quanta mass energy in packet, will be supported by S vibrations in one second by forming number of EM waves or orbitums equal to the number of Quanta mass in packet.

This inter relation ship of one unit mass (S free treos), supported by unit energy (S kinetons), spreads on unit space (S bound treo length cube) , in one unit time (S vibrations) of one second, is reflected in our day to day observations. We will observe this relation ship with few examples in light of treo model.

The mass of any body (similar to all mass energy packets of all type of photons or of all elementary particles) **will spread its load on all apex bound treos in its length of spread, and each apex bound treo will support the exerted load by S vibration in one second, and thus the full body is supported continously.**

Maintaining the same trend load exerted by a moving body is on its length of spread which will be the distance traveled by this body in one second.

(load at each apex bound treo = number of quanta mass energy in body/ distance in number of bound treos covered by this moving body in one second; As per treo model).

Each quanta of the load at each apex bound treo is supported by one orbitum (S kinetons) by its S vibrations in its one rotation in one second and thus the moving body will be supported for one second and continuosly by continous such vibrations.

For clarification note following example, **How load of body of one Kg mass moving at 90 Km per second will spread and will be supported?**

(a) This one Kg body is of = **0.85×10^{51} quanta free treos.**

   (1.58×19^{94} free treos per Kg/1.85539×10^{43} free treos in per quanta)

(b) This load of one KG mass **will spread on 25 meter in one second as it is moving at 25 meter per second,** when moving at 90 KM per hr.

   (25 meter 90000 meter/$60 \times 60 = 25$ metre)

(C) **1.52×10^{36} bound treos are present in 25 meter**

   ($1.52 \times 10^{36} = 25$ meter $\times 0.61 \times 10^{35}$ bound treos in one meter)

(D) 0.85×10^{51} quanta body on 1.52×10^{36} bound treos the body will **exert load of 0.56×10^{15} quanta per apex bound treo.**

   (0.85×10^{51} quanta/1.52×10^{36} bound treos $= 0.56 \times 10^{15}$ quanta)

(E) It will be supported by one shell, having **0.56×10^{15} orbitals (in $\sqrt{0.56} \times 10^{15}$ sub shells in each shell)**

(F) **One orbitum (S kinetons) by its one rotation in one second (by S vibrations) will support one quanta load (S free treos).**

The central load of mass energy of moving body spreads and supported in unit time of one second, is not only a distance in bound treos, but has a dimension of time as well. **Load of any moving body spreads on space matrix with time and then this divided load is supported accordingly.**

One unit mass load (about egg of flee) is the maximum load which can be supported at one bound treo, at its one unit gravitational centre at one graviton by S vibrations in one second.

But multiple units mass body or n unit mass body, exerts n^2 unit mass load at its gravitational centre and is supported by equal number of gravitons by its graviton coloumns placed in n spiral layered kinetic coloumn of third dimension.

This rotating kinetic coloumn will exert total force of n^2 gravitons at the apex of this kinetic coloumn (by one vibration in planck least time the load will be supported only for 1/S second) and by its S vibration per second it will support the load continuosly exerted by this n unit mass body at its gravitational center.

(1) Thus in third dimensional deformation any big body (e.g. asteroids or meteoroids) is supported at its gravitational center, by its single rotating kinetic coloumn of third dimension, **due to which all asteroids or meteoroids are seen rotating continuously on their axis in space.**

(2) A standing bicycle made up of multiple unit masses exerts a big load at its solitary gravitational centre, which is supported by one big spiral kinetic coloumn of third dimension **along with its big angular momentum** and therefore the bicycle has a tendency, to roll and fall.

But if bicycle is moving, its load will spread (load at each apex bound treo = number of quanta mass energy in body/ distance in number of bound treos covered by this moving body in one second; as per treo model) **on all apex bound treos in one second** and this divided load of bicycle on each apex bound treo along its spread is now supported by multiple smaller kinetic coloumns, accordingly with **lesser angular momentum,** and thus with lesser tendency to role the bicycle will not fall.

It is to be noted that higher the speed of bicycle higher will be number of supporting kinetic coloumns, but of smaller heights and with **lesser angular momentum.** This will prevent bicycle from rolling and falling and thus it makes it more stable while moving.

(3) A plane increasing its speed to take off, will spread its load and will exert less load on each apex bound treo and **with less deformation of space matrix at each point, the airplane can take off and fly by thus reducing its effective load at each point on space matrix and also helped according to Bernoulli's theorem.**

(4) **But if the body is not a confined mass e.g. a galaxy with many stars, it will spread itself along its wave by spreading its stars in spiral arms of this moving galaxy, while it's galactic centre itself remain supported by multiple unit black holes around its gravitational centre, bounded by cosmic strings.**

8. Avogadro Number[27] i.e. $6.02214076 \times 10^{23}$ (One Mole)

Graviton coloumns, at different quantum levels in process of its formation at different stages of its formation, are of different sizes12. When they accumulate in avogadro numbers

Avogadro number of graviton coloumns **($6.02214076 \times 10^{23}$ coloumns)** formed at first quantum level by load of unit electron, totals **one faraday charge,** (one mole charge)

---

27   **Avogadro Number**
  'Avogadro number' = 6.022×10^{23}
  '6.022×10^{23} 'unit electron charges' = one faraday charge
  '6.022×10^{23} hydrogen atoms' = one mole of hydrogen = one gram '6.022×10^{23} carbon atoms' = one mole of carbon = '6.022×10^{23} molecules of any gas' = one mole of gas (which occupies 22.4 litter at STP)

Partially formed graviton coloumns of Avogadro number of **hydrogen atoms,** at higher quantum level **is one mole of hydrogen** (i.e. **one–gram hydrogen**) or gram atom of hydrogen.

When partially formed graviton coloumns at still higher quantum levels formed for Avogadro number of **atoms of carbon** (made up of 12 nucleons), it makes one mole of carbon (or **12 gram carbon**).

Finally, Avogadro number of graviton coloumns of any element of any inorganic/organic compound is **one mole of this substance.**

Avogadro number of molecules of any gas (more or less to be same size of graviton coloumns for all gases) with their supporting partially formed graviton coloumns at their respective quantum level are **one mole of this gas and at S.T.P. it will occupy 22.4 liter of space.**

EXPLANATION

Avogadro number of Amu (One AMU i.e. Atomic mass unit 1/12 of one carbon atom mass ≈ one nucleon mass ≈ *1.660539* $^{-27}$ *Kg* ≈ 2.624974 × 10^{67} free treos (Mass of one proton or one nucleon expressed in free treos)

(1.660539 $^{-27}$ Kg × 6.02214076 × 10^{23} = 0.001 Kg = **one gram weight.**)

Remember that elementary particles are not point masses, but their **mass energy spreads on its Reduced Compton wave length** and are supported by its **one wave which forms at 2 × reduced Compton wave length,** which is arc of a circle with its radius equal to the reduced compton wave length. When this wave length is multiplied by pi (π) it gives **circumference of its orbit or Compton wave length of particle** (2 RC wave length× π).

This observation indicates **accumulation of fixed number or one mole** (or 6.02214×10^{23}) **of partially formed graviton coloumns of different circumferences** which are formed by different entities, at their respective different quantum levels; and all are placed in **one contracted unit space matrix at different levels of its contraction.**

If we talk about one mole of ideal gas we are talking about nothing but just about coloumns of one mole of gas molecules. **One unit space matrix** which contains one mole of any gas, **AT STP** will contract **to 24.07 liter volume.** As **graviton coloumns of different circumferences (at different quantum levels of its formation) of** one mole of gas molecules at STP contracts to same size to be accomodated **24.07 liter volume.**

One full unit space matrix is used in two dimensions to perform and to sustain all matter waves for one second. With increasing number of quanta in packet the RC wave length proportionately reduces.

As Reduced Compton wavelength × mass-energy any elemenatry particle in free treos = S^2

e.g. we will examine it with the example of one unit electron.

23.797258 × 10^{21} Bound treos RC wave length of unit electron × 1.44000857 × 10^{64} free treos in unit electron packet= 3.4 × 10^{86} = S^2 **Ground energy of second dimension.**

So, everything around our body is nothing but these orbits of graviton coloumns (complete or incomplete comprising of different number of waves) produced by deformation of space matrix in second dimension by all mass energy packets in different ways.

9. Biomass; Purposeful Digging of Space Matrix

Let me explain, up to now we observed that Sun or any other cosmic body gets its spherical shape by filling of its matter in its **self formed mould** (as the mass of sun accommodates in spherical pit produced in space matrix, by its own three dimensional spherical deformation of gravitational field of Sun by its own weight).

There is one proof of this speculation, that in our own solar system, the less dense mass of Saturn unable to be accommodated, in their three dimensional spherical pits/deformation (produced by their own gravitational fields) **spills out** in two dimensional deformation of its gravitational field (along the orbits of its satellites), which in case of planet Saturn, form **spectacular rings of Saturn.**

Similarly the **pre formed atomic orbits (moulds)** formed by nucleus of an atom is filled later on by electrons of required energy, and it thus form all 118 elements arranged in periodic table.

Similarly bio mass also generates preformed moulds (by purposeful digging in space matrix; some what similar, as roots of a tree digs in earth for its spreading) and desired matter is later on filled (docked) in these generated mould in space matrix to get desired shape of organism.

Non living matter produces load dependent deformations of space matrix, while bio mass produces deformations (to form desired shaped moulds in space matrix in which matter is filled later on to form desired shapes and for al parts of animals or plants) by purposeful digging (contraction) of space matrix.

These desired moulds are formed in space matrix by **directional and time framed energy transfers by molecules of bio–mass, with controlled force and with controlled fluctuations of applied degree of momentum; (similar to glass blower make glass objects of desired shape by controlled blow of air from his mouth).**

In photo synthesis the controlled energy is transferred by molecules in its surrounding to dig the mould in Space matrix of carbohydrate molecule or of protein molecule and desired atoms of carbon, hydrogen, oxygen or nitrogen docs in the specifically preformed moulds for each of them. The florescence pattern of flowers is maintained by first digging of pre form moulds in space matrix.

In womb, body grows by filling of desired proteins in time bound developed preformed moulds. The **information of time bound digging of space matrix is coded in their specific gene.**

E.g. Cellular clock regulates human spine development. Source; 'An in vitro human segmentation clock model derived from embryonic stem cells'.

Li fang Chu et al (https://doi.org/10.1016/j.celrep.2019.07.090)

1. The bio mass thus form all **specific shaped antibodies which fits on its targets (viruses or allergens), similar to specific pieces of** zig saw puzzel fit with other pieces according to their matching shapes, to get desired shape of an animal.

2. In drug manufacturing and drug targeting this docking system is followed.

Thus purposeful digging of space matrix (to form load depended desired deformations) by time controlled energy release (controlled momentum) by molecules in specific directions is the property of all biomass.

10. Fate of Universe

Our universe is a 'Quantum pendulum universe'. In it one life cycle, of S seconds, universe performs only one oscillation of this five–dimensional pendulum.

In this oscillation the total kinetic energy of universe is continuously decreasing while its total potential energy is increasing; due to ongoing expansion of universe, by slow but continuous and simultaneous uncurling of all curled up 'voids' of space matrix, [see page 46]. Thus, after this time of S seconds, after full expansion of universe at its **maximum potential energy state**, the magnitude of five negative dimensions in each void will become equal to the magnitude of five positive dimensions generated by bound treos. Then all bound treo placed alternatively with void will inhale each other to produce 'big crunch' and resulting in one unstable ten–dimensional universe.

Now the whole space matrix will disintegrate and this ten dimensional unstable universe will violently collapse to **break at its each point** to produce five positive dimensional treos and fully curled up five negative dimensional voids, and which is labeled as 'big bang'. The new baby

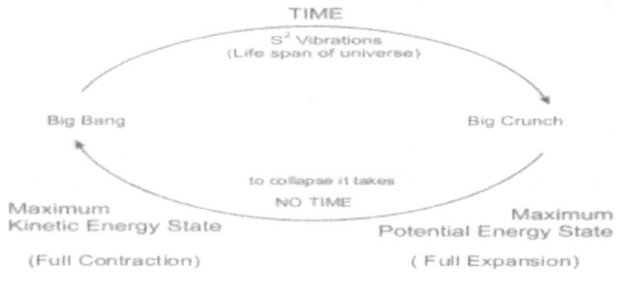

Figure 60: One life cycle of our pendulum quantum universe.

universe thus born will be at its **maximum kinetic energy state** and it will start expanding by slow uncurling of all its voids, to start the next life cycle (oscillation) of universe.

11. Multiverses (Onion Peel Model)

With increasing understanding of our universe, we are compelled to think that our universe is not a single entity, but there must be more universes which are beyond our reach. With our universe they all must have one common origin and common end.

One 'bran' in multiverse which marks our universe can be compared with one peel of onion. In our universe rhythm of its vibration is S times per second (at Planck frequency) and it decides the value of constant S (THE COSMIC CODE), which in turn decides the value of all universal constants, which regulate working of our universe.

Our universe is just one peel of an onion, is further supported by our known observation, that *our universe is flat universe, as if painted at the surface of a balloon.*

Mathematical models and this treo model support this 26-peel structure of multiverses, as if all universes are placed at their respective quantum levels (like peels of an onion) of one 'god's coloumn' according to coloumn geometry. Here it can be reminded, that all satellites of planets, all planets in our solar system, all stars in a galaxy and all galaxies in our universe are placed at different quantum levels of its parent coloumn.

In other peels or in other universes the rhythm of its vibration (as our universe vibrates at Planck frequency, by S number per second) will

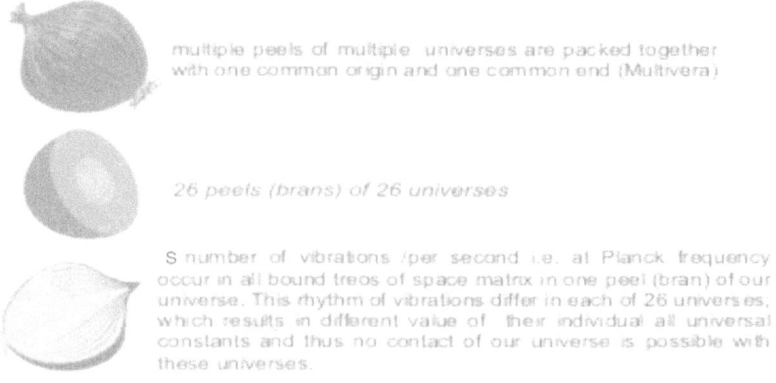

multiple peels of multiple universes are packed together with one common origin and one common end (Multivera)

26 peels (brans) of 26 universes

S number of vibrations /per second i.e. at Planck frequency occur in all bound treos of space matrix in one peel (bran) of our universe. This rhythm of vibrations differ in each of 26 universes; which results in different value of their individual all universal constants and thus no contact of our universe is possible with these universes.

Figure 61: Multiverses (proposed onion peel model)

change with change in the rate of its vibration. Thus the value of S and of all other universal constants, in every other universe will differ, with different values of their all universal constants, which will make all these universes un approachable and undetectable.

CHAPTER 5
Quantum Gravitation

1. Quantum Gravitation ('Turtles All the Way Down')

Mr. Stephen Hawkins in his book 'A Brief History of Time' has given the reference of a little old lady who after listening the lecture on gravitation, said to the orator, "Sir you may be right on some points **'but it is a turtle which supports the Earth'**. When the orator inquired 'but what supports this turtle supporting the Earth' the lady said 'smart boy, to support this turtle, there are turtles one below the other' (page 1 of book 'A Brief History of Time', Dr. Stephen Hawkins, 1988)."

I would like to point out that the old lady was right; **'infact there are turtles all the way down';** these are deformed bound treos (i.e. kinetons) or energy particles which can accumulate and generate enormous kinetic energy, as they get deformed in numbers, equal to the load of free treos they have to support. Accumulated as gravitons in the gravitational sphere they support/neutralize equal number of multiple unit masses load of cosmic body, at its gravitational center.

One graviton coloumn of each gravitons present at periphery of gravitational sphere of body, join side by side to form gravitational field of this body.

The mass pressure of cosmic body on space matrix at a distance, is supported consecutively by each individual concentric bigger layer in its gravitational field. The total kinetic energy on all sub kinetic coloumns present in any one layer of gravitational field can support total load of body on this layer. This 'diluted central mass pressure' on each increasing size of layers in gravitational field, is supported by gradually reducing deformation of space matrix in three, two and then in one dimension, till it finally fades as 'one free treo', to be supported by one kineton, (the smallest turtle of the story).

Newton calculated the value of Universal gravitational constant is the force of attraction between 'two bodies of 1 kg mass each, separated by a distance of 1 meter'. But he could not explain *how and why* **this force of attraction is generated.**

As advocated earlier; 'gravitational attraction' is not a phenomenon in itself; but it is only one effect of phenomenon of gravitation.

The bodies are supported by the forces (fields) generated due to the deformation of space matrix which **is the gravitational phenomenon,** according to the proposed model.

Two bodies will fall towards each other (named as gravitational attraction), **because of deficient kinetic support available to both bodies, by common *shared* space matrix between them. When a body, comes with in gravitational field of another body,** space matrix present between these bodies (and supplying supporting gravitons and its graviton coloumns), is shared by each other.

Both bodies share available supporting gravitons generated in this shared space matrix in proportion to their individual masses. Thus, smaller body loses the tussle and with greatly deficient number of its supporting gravitons, it falls more towards the bigger body. Even the bigger body remains deficient of some of its gravitons and it too falls a bit towards the smaller body. As the resultant forces to support both of these bodies, from the direction between them are less, than the forces from all around supporting them, so both the bodies are pushed towards each other, which is perceived as gravitational **attraction** between these two bodies.

To explain 'how', Einstein in his 'General theory of relativity' (which is a pure mathematical model) has described gravitation as 'curvature of space–time' i.e. the deformation of space produced by the mass pressure of a body (e.g. Sun). On these curved paths produced by Sun the planets move in their orbits. (dig. 8 on page 63)

In fact, his calculations depict *Schwarzschild matrix*, which describes gravitational field strength in second dimensional deformation of space matrix at which the planets are placed. He said that the *Earth and other planets move straight, but it is the path which is curved, due to presence of Sun. The difference in 'Slope of deformation' produced by each body, results in fall of bodies towards each other, to which we notice as gravitational attraction.*

Einstein while describing gravitation as deformation of space–Time has not described the mechanism and purpose (i.e. Why) of this deformation of space–time. Thus, we see that both above (although right) descriptions of gravitation, remains incomplete while describing the force.

Gravitation can be described as a reaction by the space matrix to neutralize the mass pressure of a body which may range from photon to unit black hole. Thus, the space matrix in response to 'mass pressure of Sun' generates a 'kinetic energy pool' around it (gravitational centre of Sun) and to neutralize its mass pressure at a distance the gravitational fields are formed. After thus supported, the Sun gets converted into a weightless symbolic 'point mass' on space matrix.

It is for this reason that when Galileo, simultaneously dropped balls of different masses from 'leaning tower of Pisa' they all reached the ground at the same time. The 'mass pressure' of all these different size balls were neutralized by their individual gravitational fields and now 'weight less balls', after being converted into 'symbolic point masses', were affected only by the (legend) gravitational field of Earth and they all came down to the ground together at the same time.

Another example of this universal behavior of all bodies, to present themselves as 'symbolic weight less point masses' on space matrix, is confirmed in planetary motion.

The 'speed of revolution of all planets' does not depend upon their individual different masses, but only depend on their distance from Sun.

The individual masses of all cosmic bodies are neutralized at their gravitational centre by their gravitational spheres, while the diluted mass pressure exerted on each layer and at each point on surrounding space matrix is neutralized by gravitational kinetic energy generated at this point in their individual gravitational fields.

Now all 'weight less point masses of all these planets on space matrix' revolve at the speed of v bound treo distance per second as 'symbolic point masses', only affected by gravitational kinetic energy of sun (which can be calculated at any point in gravitational field as v^2 equal to Mg/r) present in their orbits.

2. Formation of Gravitational Field

(a) **The load of body exerted at the apex (top) of one kinetic coloumn is distributed on its layers according to this coloumn geometry (Fig. 14, page 87)**

To understand the dispersal of load in a kinetic coloumn, we will take the example of 9 layered kinetic coloumn, which support 81 (or 9^2) free treos load at its apex.

81 'free Treos' 'load' is supported by equal number of 81 'kinetons' present in this 9 layered kinetic coloumn.

The total load exerted at apex of kinetic coloumn distributes equally on all kinetons, up to any layer in coloumn which is taken in account.

If you take account up to **second layer,** the load divides as ¼ on all three kinetons (diluted mass pressure 'a') of second layers (as total 1+ 3 = 4 kinetons are in coloumn), while one kineton of first layer supports full load.

When third layer of sub kinetic coloumn is taken in to account, it is $1/9^{th}$ of total load (diluted mass pressure 'a') on each of 5 kinetons of third layer, as there are total nine bound treos up to first three layers of this coloumn, (while it is ¼ of total load on each kineton in second layer).

It distributes as $1/16^{th}$ of total load (diluted mass pressure 'a') on each of 7 kinetons in **fourth layer** as there are total '16 bound treos to support the load up to fourth layer in coloumn', while it exerts 1/9 of load on each 5 kinetons of third layer and ¼ of total load on 3 kinetons of second layer.

The same pattern of K, L, M, N energy distribution of atomic energy in atomic orbits is seen where, K (E energy), L (E/4 energy), M (energy E/9), N energy levels (energy E/16).

Similarly if we calculate the **load in ninth layer** of coloumn as it divides on all 81 bound treos (kinetons) in kinetic coloumn and it is 1/81 at each of 17 kinetons (2n-1) in outermost ninth layer. (See Fig 14 page 87)

The above mentioned load distribution pattern of any load at its apex, calculates the diluted mass pressure on any one kineton, which is **'a'** **diluted mass pressure of body in treo model** according to this coloumn geometry.

This can also be calculated by Newton's formula 'a'= MG/r^2.

Where 'a' is diluted mass pressure and = M (or central load)/r^2.

Here M is total load at apex of kinetic coloumn = 81 free treos load, **while at a distance of ninth layer (when r = 9) it is calculated as 'a' = $M/9^2$ or = M free treos/81**

But, traditionally 'a' is acceleration (where 'a' or Newton's acceleration is calculated by same formula, a = MG/r^2). **Thus 'a' of Newton's acceleration is diluted mass pressure 'a' in treo model.**

If Central load ----- If Load of 81 free treos is M, as it is supported at apex of 9 layered or 'r' layered kinetic coloumn which have = r^2 **or 9^2 free treos as per coloumn geometry.**

Calculation of Diluted mass pressure – – Then diluted mass pressure 'a' exerted by this load on each bound treo (kineton) of ninth layer in this kinetic coloumn is calculated as **'a' = M free treos/r^2 or 9^2**

Calculation of Load at one bound treo – – Now if **we calculate (r × a) load** at each bound treo in this outermost ninth layer of sub kinetic coloumn it is = **9 (r) × M/9^2 (a) = M/9.**

Calculation of Load on n or all 9 bound treos – – – – Total load M of body spreads on 9 bound treo = 9 × ra = 9 × M/9 = M. (on RC wave length)

Calculation of Total exerted load of 2 M on 2n-1 bound treos, which form one layer of kinetic coloumn, (on which one wave form with time)

Total load on 2n-1 or on 17 bound treos [2 × 9 – 1= 17].

On 17 bound treos of this layer the 2M total load of central mass is distributed and supported. It is (17 × M/9) = 2M – 1 = 2M, as 1 is negligible when other numbers are big.

(b) **This coloumn geometry (load distribution pattern) also decides size of all matter waves.**

Length of spread of any packet = on its RC wave length in bound treo distance = S bound treos/total number of quanta in packet.

If 'n' is reduced Compton wave length (where 'n' is half of the length of its wave) **at any n^{th} quantum level at which mass energy of packet will spread,** while one full wave is formed on one full layer of kinetic coloumn on **2n–1 kinetons present in this layer** ($\approx$ in $1/3^{rd}$ arc of a circle of orbit).

And (2n–1) × π is angular momentum of this mass energy packet or its **Compton wave length or full circumference of orbit**. (Fig.62)

This relation we see when the total mass energy of **any elementary particle** at any distance r or n; is supported at n apex bound treos (on half length of any one layer of kinetic coloumn), while 2n-1 apex bound treo in one layer forms its wave, and circumference of its orbit 2n × π.

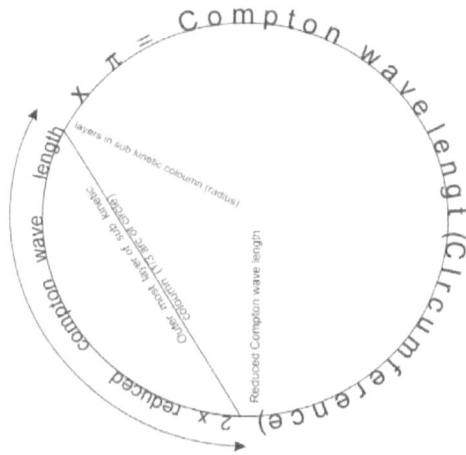

'n' bound treos = radius of circle =Reduced compton wave length
2n-1 bound treos = one matter wave = n^th One layer of coloumn
2n-1bound treos x π = circumference of circle = Compton wave length

Figure 62: Outer most layer of n layered coloumn has 2n-1 kinetons on which one wave form. Here n is reduced Compton wave length, while (2n-1) × π is Compton wave length or circumference of circle.

(c) Distribution of central load of body in gravitational field

The same we see in gravitational field; any M unit mass body, made up of MG free treos exerts a load of MG^2 free treos at its gravitational centre which spreads as 2MG in any one direction and in one plane, on any one layer (on 2 n–1 bound treos) of gravitational (coloumn) field.

We can calculate 'a' or diluted mass pressure of this MG load at any of 2n-1 apex bound treo in any layer at distance r, by Newton's formula $a = MG/r^2$; or by coloumn geometry as described above.

Then at each apex bound treo of any layer the load $r \times a$ can be calculated.

The **2 MG mass pressure of body** exerted on any layer is evenly distributed, on each of $2n-1$ apex bound treos as 'r a' in n^{th} layer of gravitational field.

But If n = r, (distance of this orbit from sun in number of bound treos), then this total load of **2MG free treos** is on all 2r-1 apex bound treos which form a wave.

This load at each apex bound treo is supported by one sub kinetic coloumn while all kinetons in all sub kinetic coloumns (one on each apex bound treo) in one full layer has **2MG kinetons.**

Thus this gravitational kinetic energy in one layer which supports/ neutralize 2 MG load of free treos comes from $2n-1$ sub kinetic coloumns which forms at each apex bound treo in this one layer. All kinetic coloumns have (equal number) total **2MG kinetons;** which are equal to $v^2(2r-1)$, present in all $2r-1$ kinetic coloumns in this n^{th} layer from Sun.

At distance r the reacting kinetic energy on each apex bound treo v^2 can be calculatd by Newton's formula $v^2 = MG/r$ or by coloumn geometry as described above.

So, it can be seen that **Newton's equations which describe gravitational field are in accordance with this proposed coloumn geometry.**

Newton's gravitational field equations in fact describe coloumn geometry: Description of the gravitational field of Sun given by Newton is the description of these fields in one dimension and in one plain only; but actually, **the gravitation is four–dimensional deformation**

of space matrix, which is required to generate the kinetic energy at gravitational center of Sun to support it.

All fields are made up of kinetic coloumns and the coloumn geometry obeys 'inverse square law'. The gravitational field, intensity of light, charge density etc all fade *'by reciprocal of the square of distance 'r' '*.

Quantum gravity has positive energy [ref. 5]. **Quantum gravitation in its most basic form can be defined as the reaction of space–matrix @ one free treo is supported by one kineton (one deformed bound treo) by which it neutralizes the mass–pressure of a body, where the body may range from photon to multiple black holes at galactic center.**

Unit mass or Planck mass is the maximum load which can be supported at one point i.e. at one bound treo on its unit gravitational center with the formation of one graviton and its graviton coloumn.

All type of packets, which have *mass energy less than one–unit mass*, spread this mass energy on all apex bound treos along its RC wave length **(S bound treos/Number of quanta mass energy in packet) and infact are not particles but all are waves.**

The photon packet is supported by all sub–kinetic coloumns present at each apex bound treo along its wave length which together form **one EM wave.**

While each individual rotating kinetic coloumn (shell) present at each apex bound treo along Reduced Compton wave length of any elementary particles packet, together form **one matter wave.**

But for the bodies made up of multiple unit masses, any M unit mass body, exert a load of square number of unit masses present in this body i.e. M^2, at its gravitational center, and then this each unit mass load at gravitational centre is supported by one graviton (and its graviton coloumn) present in 'electron black hole' in third and in 'gravitational sphere' in fourth dimension.

All of these gravitons are present in one **electron black hole** in third dimension and can support up to one billion metric ton mass body.

The central load of cosmic bodies is supported by four dimensional deformations, at their individual gravitational center by all gravitons present in gravitational sphere of body.

Afterwards this central load spreads outside this gravitational sphere, all around on space–matrix and exerts its 2MG mass pressure of central load on each concentric layer of gravitational (coloumn) field.

The central mass pressure (according to its diluted mass pressure), is first supported by three–dimensional deformation (which accommodates and shapes all spherical cosmic bodies) and then by two–dimensional deformation, where at its fixed quantum levels planetary orbits are formed, in which baby bodies (planets, satellites and rings formed by debris, gas and dust e.g. Ring of Saturn) condense and revolve in their respective matter waves. In gravitational field of Sun after the end of this two dimensional deformation, the curve or slope of bowel shaped deformation of sun will end.

Finally, in one dimensional deformation according to proposed coloumn geometry, the flat gravitational field extends up to its acceleration zone (till the last layer of gravitational field of body, where 1 free treo load is supported by one kineton) (Fig 63)

3. Structure of Gravitational Field

Gravitational (coloumn) field is formed by **union of all graviton coloumns, one on each graviton present in outer most layer of gravitational sphere (in any one direction and plain).**

Thus the n bound treo layered gravitational sphere of n unit mass cosmic body will have 2n-1 gravitons in its outermost layer, and 2n-1 graviton coloumns formed on these gravitons will join together (side by side) to form gravitational field of this cosmic body.

M unit mass of body exert a load of M^2 at its gravitational center, where it is supported by M^2 number of gravitons in its one gravitational sphere.

While, its mass pressure in any one direction and on any one layer of its gravitational field (coloumn) is of **2M free treos**, which is supported by total **2M kinetons** as gravitational kinetic energy, [present in all kinetic coloumns at $(2n-1)$ apex bound treos of any n^{th} layer of gravitational field].

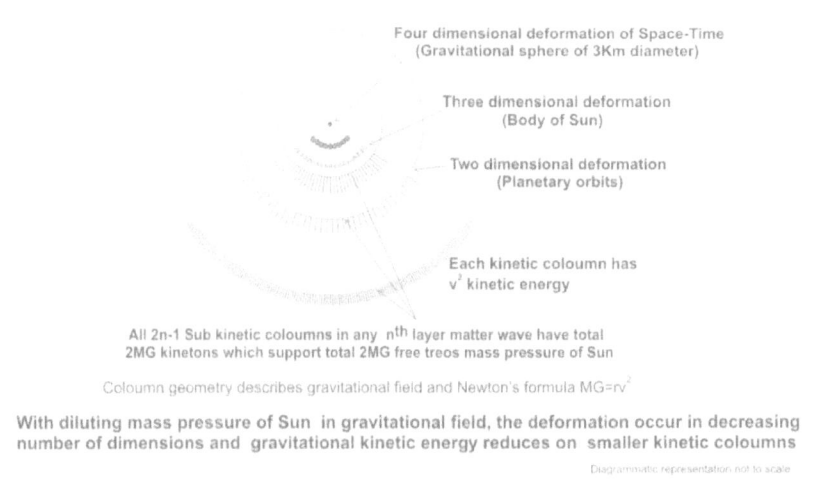

Four dimensional deformation of Space-Time
(Gravitational sphere of 3Km diameter)

Three dimensional deformation
(Body of Sun)

Two dimensional deformation
(Planetary orbits)

Each kinetic coloumn has
v^2 kinetic energy

All 2n-1 Sub kinetic coloumns in any n^{th} layer matter wave have total
2MG kinetons which support total 2MG free treos mass pressure of Sun

Coloumn geometry describes gravitational field and Newton's formula $MG=rv^2$

With diluting mass pressure of Sun in gravitational field, the deformation occur in decreasing number of dimensions and gravitational kinetic energy reduces on smaller kinetic coloumns

Diagrammatic representation not to scale

Figure 63: Gravitational Sphere and Gravitational Field of Sun

The 2 MG mass pressure of body exerted on any layer is evenly distributed, on all 2n-1 apex bound treos in n^{th} layer of gravitational field.

But, if n =r, distance of this orbit from sun in number of bound treos, then this total load of **2MG free treos** which is equal to r × a (2r-1).

It is supported/neutralized by gravitational kinetic energy of equal number of total **2MG kinetons** which are equal to v^2 (2r−1), present in all 2r−1 kinetic coloumns in this n^{th} layer from Sun.

Here n = r is number of bound treo layers from sun up to orbit; 'a' is diluted mass pressure (a = MG/r^2) of central load at this bound treo in orbit; r × a is load which is supported at this bound treo by v^2 kinetons in one v layered kinetic coloumn. This (ACTION) ra free treos = (REACTION) v^2 kinetons present in each kinetic coloumn.

2 MG free treo load = Dilutes on all apex bound treos in orbit × ra

= 2 MG kinetons = in all kinetic coloumns placed on all apex bound treos in orbit × v^2

Thus, the total load of 2MG of body on any one layer is supported by 2MG gravitational kinetic energy in each successive bigger single layer of its gravitational field. (Fig. 63)

Figure 64: gravitational kinetic energy 2MG in each layer, concentrates gradually (on lesser number of apex bound treos and more layered kinetic coloumn) in smaller concentric orbits in gravitational field towards sun, as depicted in Schwarzschild matrix of general theory (i.e. gravitational field of Sun)

4. The Matter Wave in any n^{th} Gravitational Field Layer From Sun

a. While total load supported by one wave at any n^{th} layer is **(2n−1) × ra = 2MG.**

At any one, out of **2n−1** apex bound treos (forming its matter wave) in n^{th} layer, the action r × a **(load of free treos) at each apex bound treo,** is reacted by equal force by v^2 kinetons, in one v layered sub–kinetic coloumn.

b.   At each apex bound treo, $\mathbf{r\,a = v^2}$

c.   Analogous equation that we get in general theory of gravity is

$$G = 8\,\pi\,G/c^4\,T$$

This 'ra' or load, is analogues to T (mass pressure, or **energy momentum tensor**), while G in general theory equation is **Einstein's tensor** which measures curvature of space and generates v^2 or equal gravitational kinetic energy (by equal reaction) at this point.

d.   **While total gravitational kinetic energy in wave at any n^{th} layer is $(2n-1) \times v^2 = 2MG$**

In these equations;

r or n = is distance of nth bound treo layer of gravitational field of sun from centre of Sun; and on 2n–1 apex bound treos of this layer one matter wave of this orbit is formed.

a = Newton's acceleration and in this model, it is 'diluted mass pressure of Sun' at each apex bound treo (and it can be calculated by Newton's formula, $a = MG/r^2$, or by proposed coloumn geometry).

v^2 = Number of kinetons in any v layered sub–kinetic coloumn which is present at each apex bound treo of matter wave (which can be calculated with its total number of kinetons by Newton's formula $v^2 = MG/r$, or by proposed column geometry)

v = Number of bound treo layers in any one sub–kinetic coloumn = orbital speed of planet placed in this orbit in terms of number of bound treo distance per sec = frequency of matter wave = number of quanta load at this orbit.

MG = **Number of free treos** as mass energy in body of Sun = Mass of Sun in terms of unit masses × Quantized value of gravitational constant

Conventional value of Mass of Sun in kg × conventional value of Gravitational constant.

5. New Derived Value of Gravitational Constant

(After Replacing Meter with Planck Least Length)

Gravitational constant 'G' = 6.67430 × 10⁻¹¹ Meter³ per kg, per sec per sec. (Dimensional Formula $L^3 M^{-1} T^{-2}$)

a. By substituting the value of meter by natural unit of Planck's least length **'G' can be recalculated** (while **one meter have 0.61871425 × 10³⁵ bound treos** and each can be converted in one kineton)

   $6.67430 × 10^{-11} × {}^{28}(0.61871425 × 10^{35})^3$ per kg, per sec per sec

   $1.58079692 × 10^{94}$ kinetons per kg, per second per second.

It also means that this number of kinetons act on 1 kg mass per second per second and will support equal number of 'free treos' in one KG. Thus $1.58079692 × 10^{94}$ 'free treos' constitute one kg mass.

b. **Alternatively, the value of G can also be denoted by number of kinetons supporting one–unit mass (S^2number of free treos)**

   $3.442488398 × 10^{86}$ kinetons per unit mass, per second per second.[29]

c. **Alternatively, the value of G can also be denoted by kinetons acting on one free treo (as in case of unit photon)**

   1 kineton per free treo, per sec per second.

---

28   One meter = 1/Planck Least Length = $0.61871425 × 10^{35}$ bound treos distance.

29   As one–unit mass is maximum load which can be supported at 'one – unit gravitational center', (also named as graviton) in universe, this derived new unit for one unit mass is fundamental and true unit of gravitational constant.

Then gravitation as a phenomenon can be explained as

G = universal constant = 1unit action by one free treo reacted by one kineton, per second per second (i.e. continuously)

6. What Gives Slope to Gravitational Field?

Increasing gravitational kinetic energy (v^2) in each kinetic coloumn present at all apex bound treos in each next layer towards Sun, is 2MG kinetic energy. **But this 2 MG kinetic energy does not remain constant but comparatively reduces in each successive layer (matter wave) towards Sun, as the value of factor −1 increase disproportionately fast.**

This happens with decreasing value of n (as n is also equal to radius or distance from sun) in equation [2MG >= (2n−1) × v^2] the impact of −1 factor increases disproportionately fast as n approaches 1.

$$2MG >= (2n-1) \times v^2 >= MG$$

(r or n is large & 1 is small)

In each successive layer towards Sun, the less value of 2MG is calculated (i.e. total gravitational kinetic energy in one layer) than in next outer layer according to formula 2MG >= (2n−1) v^2.

The hypothetical calculation of 'total kinetic energy in one layer' from layer number 1 to layer number 10 in gravitational fields **(is calculated as 1 MG, 1.5 MG, 1.66 MG, 1.75 MG, 1.80 MG, 1.83 M, 1.85 MG, 1.87MG, 1.88 MG and 1.9 MG)**. This anomaly provides a gentle slope towards parent body and thus the bodies in gravitational field, due to the total force of one full outer bigger layer, will slide/ fall towards Sun.

7. Slowdown of Time as Gravitational Field Strength Increases Towards Sun

Time slows down in gravitational sphere of Sun, due to reduction of one vibration per quantum level per second, as we move towards gravitational center and the same is reflected at each corresponding √S gravitational quantum levels with more and more contracted space in gravitational field towards Sun.

This slow down of time, is required to gradually reduce the speed of rotation of each cocentric layers of gravitational sphere towards centre, so that each inner quantum layer of reducing size (of $2n-1$ electron black hole) of this kinetic coloumn of gravitational sphere placed at succesively reducing radius, could rotate simultaneously. And thus whole coloumn maintaining its shape in $1/3$ area of gravitational sphere (in $1/3$ arc of a circle) can exert its total force (of all n^2 gravitons in all 'n' bound treo layers of this gravitational coloumn) jointly at its gravitational centre, as one unit to neutralize n^2 unit masses load at gravitational centre of this n unit masses body, by its S vibrations while rotating in all S planes (360^0) in one second.

8. Gravitational Attraction

The body is supported by its gravitational field from all sides, but when another body comes in its gravitational field; **both bodies share one common space matrix in between.** Now both bodies get inadequate support

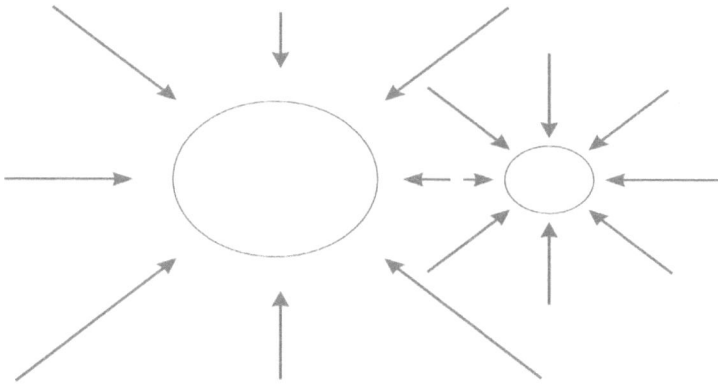

Figure 65: All bodies are pushed from all possible sides for being supported; once another body comes in gravitational field of any body, both bodies get inadequate support (from common shared space matrix in between), and then both fall towards each other, and we perceive this fall of body as gravitational attraction of other body.

From this shared matrix in between and they fall towards each other.

Gravitational attraction is **fall of bodies towards each other due to the deficient support to bodies, provided by common shared space matrix in between. The number of supporting gravitons formed in, this common shared space matrix in between, are divided in proportion to their individual weight of both bodies, but remains inadequate for both bodies to be supported. (Fig.65)**

CHAPTER 6

Planets and Satellites

1. Formation of 'Gravitational Quantum Levels' in Gravitational Field of Sun (Gravitational Waves)

As 10^{38} unit masses are in the body of Sun, it forms a gravitational sphere of 10^{38} bound treo layers to support it.

In gravitational field of Sun, outside this gravitational sphere, for first quantum level 1×10^{38} **bound treo layers** of gravitational field are required, for second quantum level **next 3×10^{38} bound treo layers** are added, at third quantum level next 5×10^{38} **bound treo layers** are added, for fourth quantum level 7×10^{38} **bound treo layers are added** in radius and according to coloumn geometry, **subsequently $2n-1 \times 10^{38}$ bound treo layers** are added at any next n^{th} quantum levels.

Thus, bound treo layers up to any desired quantum level = **Square of quantum number × Bound treo layers in gravitational sphere of parent body** (which are also equal to the Number of unit masses, as mass energy of parent body)

2. Number of Bound Treo Layers, From Its Gravitational Center of Sun to $10^{4\,th}$ Quantum Level

In second dimensional deformation in any gravitational field, at $10^{4\,th}$ quantum level the baby bodies (Planets and Satellites) condense.

Now if we calculate total number of bound treo layers (the distance from gravitational center of Sun) **up to any n^{th} quantum level it will be total $n^2 \times 10^{38}$ bound treo layers in gravitational sphere.**

Distance of 10^{4th} quantum level from center of Sun, when calculated in number of bound treo layers = square of quantum level $\times 10^{38}$ bound treo layers in gravitational sphere of sun = $(10^4)^2 \times 10^{38} = 10^{46}$ bound treo layers in gravitational coloumn of Sun from its gravitational center.

3. Planets are Placed at 10 Energy Levels From 10^{4th} to 10^{5th} Quantum Level

We find the all big solar planets in second dimensional deformation of gravitational field of Sun, and are placed at ten energy levels in 10 planetary quantum energy levels, between 10^{4th} & 10^{5th} gravitational field quantum levels.

4. Calculation of Gravitational Kinetic Energy in Orbit of Mercury 'E'

E = total number of kinetons (v^2) in any one 'sub kinetic coloumn', at this 10^{4th} quantum level in gravitational coloumn of sun, at which lies the planetary orbit of our first planet Mercury' (according to *formula* $v^2 = MG/r$)

The Kinetic energy in each sub kinetic coloumn in orbit of Mercury

Em = 8.776444393 $\times 10^{78}$ kinetons = v^2 = Mg/r ($3.145233887 \times 10^{124}$ /$0.3583722239 \times 10^{46}$)

5. Position of Planetary Orbits at Different Quantum Levels of Sun

Name of planet	E (gravitational Kinetic energy in one kinetic coloumn in its orbit)	Calculated	Actual
Mercury	E	$8.776444393 \times 10^{78}$	$8.776444393 \times 10^{78}$
Venus	E/2	$4.38822197 \times 10^{78}$	$4.696829258 \times 10^{78}$
Earth	E/3	$2.925481464 \times 10^{78}$	$3.397352232 \times 10^{78}$
Mars	$E/4(E/2^2)$	$2.194111098 \times 10^{78}$	$2.230118005 \times 10^{78}$
Asteroid belt	$E/9(E/3^2)$	$0.975160488 \times 10^{78}$	
Jupiter	$E/16(E/4^2)$	$0.548527774 \times 10^{78}$	$0.653301797 \times 10^{78}$
Saturn	$E/25(E/5^2)$	$0.351057775 \times 10^{78}$	$0.356398183 \times 10^{78}$
Centures I	$E/36(E/6^2)$		
Uranus	$E/49(E/7^2)$	0.1791111×10^{78}	$0.177026782 \times 10^{78}$
Centures II	$E/64(E/8^2)$		
Neptune	$E/81(E/9^2)$	$0.108351165 \times 10^{78}$	$0.113015946 \times 10^{78}$
Kupier belt object Pluto	$E/100(E/10^2)$	$0.087764443 \times 10^{78}$	0.0860632×10^{78}

$E/10^2$

Plutinos– Orcus 90448(2004 DW), Luxion (2001 KX76), 2202 Tx300 (55636), 2003 EL61 (136108), Quaor 2002 LM60 (50000)

$E/11^2$

Cubewanos– 2005 FY9 (136472), 2002 AW197

Scattered disc objects—

E/12² 2002 Tc$_{302}$, 2001 XR$_{190}$, 2002 YW$_{134}$

E/13² Eris (2003 Ub$_{313}$) Tenth planet

E/14² 2005 TB$_{190}$, 1996 TL$_{66}$ (15874),

E/15² 2001 FZ$_{173}$, 1996 GQ$_{21}$,

E/16² 2003 FX$_{128}$,

E/17² 2004 PB$_{112}$,

E/18² 1999 RD$_{215}$, 2000 PJ$_{30}$,

E/21² 2005 PU$_{21}$, 2003 HB$_{57}$,

E/24² 2001 FP$_{185}$,

E/30² 2004 VN$_{112}$

E/35² Red Planetoid **Sadna** 2003 VB$_{12}$,

E/40² 2002 OO$_{67}$,

E/49² 2006 Sq$_{37}$

Table 3: Planets and Planetoids are placed at planetary quantum levels.

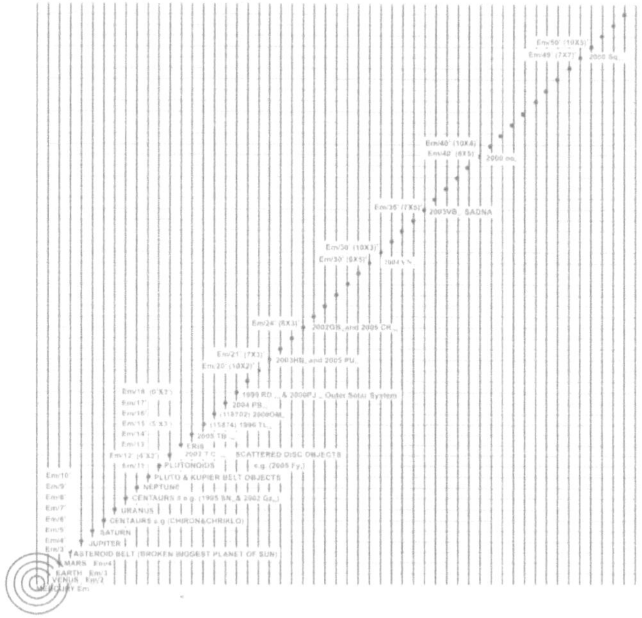

Energy level in the orbit of planets as calculated by the energy
level in resept to the orbit of mercury

Figure 66: Graph of gravitational kinetic energy in gravitational field of Sun

In Sub shells at first planetary quantum level

E Kinetons are present in each 'sub kinetic coloumn' in the orbit of
Planet Mercury in *first sub shell* of first planetary quantum level.

(**E/2**) kinetons are present in each 'sub kinetic coloumn' in the orbit
of **Planet Venus** in *second sub shell* of first planetary quantum level.

(**E/3**) kinetons energy are present in each 'sub kinetic coloumn' in the
orbit of **Planet Earth** in *third sub shell* of first planetary quantum level

The gravitational kinetic energy 'E' reduces by '**square of planetary
quantum number,** at which the planet is placed'.

E/4 (E/2^2) at Planet Mars,

E/9 (E/3^2) at Asteroid belt,

E/16 (E/4²) at Planet Jupiter,

E/25 (E/5²) at Planet Saturn,

E/49 (E/7²) at Planet Uranus,

E/81 (E/9²) at Planet Neptune,

and E/100 (E/10²) at Pluto.

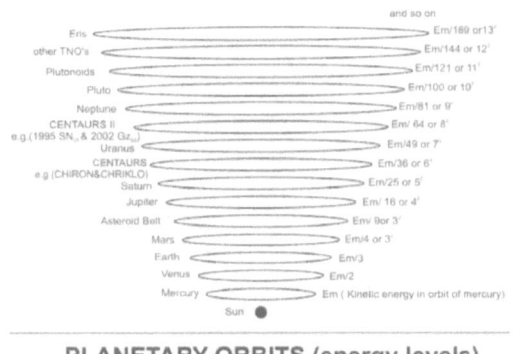

PLANETARY ORBITS (energy levels)

Figure 67: Planets are placed at planetary quantum levels

6. Energy Levels in Orbit of Planets

All the planets condense in 10 planetary quantum levels In any gravitational field between 10⁴th and 10⁵th quantum level of fourth dimension.

Gravitational kinetic energy Em reduces at 10 Planetary quantum levels in the planetary orbits by Em /square of planetary quantum level number. (Here Em is energy in the orbit of mercury).

At 10⁴th quantum level from Sun, first planetary quantum number 'n1'; in sub shells of one shell, Mercury, Venus, Earth are placed at K, L, M energy levels. While Mars is at 'N' energy level, which can also be leveled as 'n2' or second planetary quantum level **'n2'**.

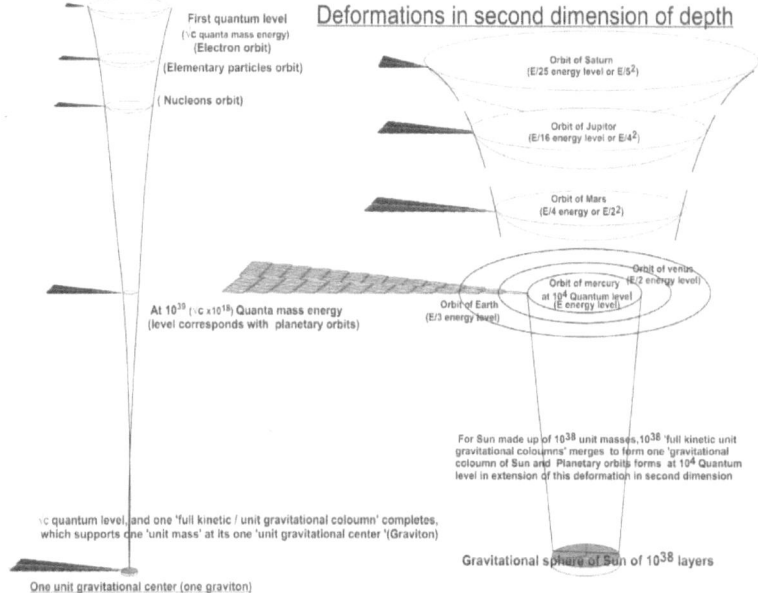

Figure 68: 10^{38} graviton coloumn together unite side by side to form gravitational field of Sun and at identical quantum level they unite to form planetary orbit.

First planetary quantum number 1 'n1' level (in first sub shell), is at unit distance of 0.3×10^{46} bound treo layers from gravitational center of Sun (at K energy level of first quantum level) at which planet Mercury is placed, with gravitational kinetic energy E.

First planetary quantum number 1 (in second sub shell) at double the unit distance ($2 \times 0.3 \times 10^{46}$ bound treo layers) and $E/2$ energy (at L energy level) the planet Venus is situated.

First planetary quantum number 1 (in third sub shell), at three times the unit distance ($3 \times 0.3 \times 10^{46}$ bound treo layers) and $E/3$ energy (M energy level) the planet **Earth** is placed.

Similarly, at **planetary quantum level 2,'n2'**, at $E/(2)^2$ i.e. four times the unit distance and at $E/4$ energy level the planet **Mars** is placed. (N energy level)

At **planetary quantum level 3,'n3'**, at $E/(3)^2$ i.e. 9 times the unit distance and E/9 energy level the **'Asteroid belts'**;

At **planetary quantum level 4,'n4'**, at $E/(4)^2$ i.e.16 times of unit distance the **Jupiter** with E/16 energy level;

At **planetary quantum level 5, 'n5'**, at $E/(5)^2$ i.e. 25 times the unit distance the **Saturn** with E/25 energy level and so on;

At **planetary quantum level 6,'n6'**, at $E/(6)^2$ i.e. 36 times of unit distance the 1977 UB (CHIRON)/and 'CHRIKLO 1997 CU26 with E/36 energy level;

At **planetary quantum level 7, 'n7'**, at $E/(7)^2$ i.e. 49 times of unit distance away the **Uranus** is placed with E/49 energy level.

At **planetary quantum level 8, 'n8'**, at $E/(8)^2$ i.e. 64 times of unit distance the planetoid "1995 SN 55" is placed with E/64 energy level.

At **planetary quantum level 9, 'n9'**, at $E/(9)^2$ i.e. at 81 times of unit distance away the **Neptune** with E/81 times energy level is placed.

While **Pluto** demoted from a full planet, is at **planetary quantum level 10, 'n10'**, at $E/(10)^2$ i.e. 100 times of unit distance is placed. The Pluto with exactly E/100 energy level (and thus it compels us to re –consider the Pluto as a full planet instead of its demoted status of minor planet)

At **10^{5th} quantum level from Sun** $[(10^5)^2 \times 10^{38}$ layers] in 10^{48} th bound treo layers in gravitational field of Sun the outer solar system exists.

Planetoids are placed at 18th planetary quantum level, **'n18'** i.e. **18th** (or 3×6) level and also seen at, **'n21' 21st** (or 3×7), and others are at **35th** (or 5×7), **24th** (or 3×8), **40th** (or 5×8), and at **49th** (7×7) planetary quantum levels. **It leaves following positions un filled and it compel us to search (The Planetoid which are at 3×5 i.e. fifteenth and at 5×6 i.e. thirtieth planetary quantum level)**

7. Earth Orbit

Mass of Sun is $0.91379754 \times 10^{38}$ Unit masses. The one gravitational field (coloumn) of Sun forms by merging of $2 \times 0.91379754 \times 10^{38} - 1$ full graviton coloumns (2n-1); of $2 \times 0.91379754 \times 10^{38} - 1$ gravitons present in outer most n^{th} layer of its gravitational sphere (of 3 Km diameter, by this model and also by general theory of relativity).

(A) Earth is situated at $\mathbf{1.00658064 \times 10^{4\,th}}$ **'gravitational quantum level'** of Sun.

Distance of orbit of planet Earth in bound treos layers from Sun = **Square of this quantum level × number of layers in gravitational sphere of Sun.**

At square of $1.00658064 \times 10^{4\,th}$ 'gravitational quantum level' of planet Earth (or 1.01320458×10^{8}) × when multiplied by $0.91379754 \times 10^{38}$ layers in gravitational sphere of Sun = $\mathbf{0.925863862 \times 10^{46}}$ **bound treos layers from Sun, is the distance of orbit at which planet Earth is situated.** (This distance calculated in bound treos when converted in Km; we get the conventional distance of earth in Km from sun, which again proves treo model)

At this level of Earth orbit, the **Reduced Compton wave length of Sun is also = $0.925863862 \times 10^{46}$ apex bound treos,** on which M load of sun spreads on RC wave length, and is supported by action – reaction mechanism.

(B) load of Sun, at any one apex bound treo, in this Earth orbit is neutralised by equal number of kinetons in one kinetic coloumn

1. 00658064×10^{4} gravitational quantum level of Sun, corresponds to $1.84319077 \times 10^{39}$ quantum level of quantum world with equal **$1.84319077 \times 10^{39}$ layers in both sub kinetic coloumns**

   a. $0.91379754 \times 10^{38}$ fully formed graviton coloumns togather form gravitational (coloumn) field of Sun. The load ra (of diluted mass pressure a) **3.3999573 × 10⁷⁸ free treos** is supported by **1.84319077 × 10³⁹ layered kinetic coloumns made up of 3.3999573 × 10⁷⁸ kinetons at 1.00658064 × 10⁴ quantum level at a distance of (1.00658064×10⁴)² bound treo layer when counted from below i.e. from apex of any one graviton coloumn.**

In one fully formed graviton coloumns it will have $1.84319077 \times 10^{39}$ quanta load and **1.84319077×10³⁹ layered sub kinetic coloumns at 1.00658064×10⁴ quantum level (1.84319077×10³⁹** = 1.85539 × 10 ⁴³/1.00658064×10⁴ quantum level) and ra load.

This ra Load can also be calculated as per coloumn geometry (a diluted mass pressure × r) of **3.3999573 × 10⁷⁸ free treos (3.3999573 × 10⁷⁸** = 3.44247205 x10⁸⁶ free treos in one unit mass supported by one graviton coloumn/**(1.00658064×10⁴)² bound treo layer from apex:** load = MG/r²)

b. Same mass of $1.84319077 \times 10^{39}$ quanta, at $1.84319077 \times 10^{39}$ quantum level during its formation (quantum level is counted from above) forms $1.84319077 \times 10^{39}$ layered sub kinetic coloumns having 3.3999573×10^{78} kinetons $(1.84319077 \times 10^{39})^2$ to support a load of 3.3999573×10^{78} free treos in quantum world.

2. ACTION REACTION MECHANISM (ACTION = REACTION)

ACTION (r x a) = load 3.3999573×10^{78} free treos at each apex bound treo.

REACTION (v²) = 3.3999573×10^{78} kinetons in each kinetic coloumns which form at each apex bound treo.

a. This load **3.3999573 × 10⁷⁸ free treos load is action ra.** ('r × a' diluted mass pressure of Sun 'a' × at distance 'r') is an action at any one apex bound treo in orbit of Earth.

3.3999573×10^{78} free treos load at each apex bound treo = r × a.

Where '**r**' = **0.9258638622 × 10⁴⁶ Bound treo layer** distance of earth orbit from Sun.

And '**a**' = **3.67 × 10³² free treos** diluted mass pressure of Sun at earth orbit.

There fore r × a = $0.9258638622 \times 10^{46} \times 3.67 \times 10^{32}$ = **3.3999573 × 10⁷⁸ free treo load.**

(Here both 'r' and 'a' are calculated in terms of treos, are equal to conventional values when calculated in SI units on conversion.

NOTE DOWN; $a = MG/r^2$ *is acceleration according to Newton's equations,* and same is calculated here as diluted mass pressure of Sun at each apex bound treo.)

b. This load 'ra' at each apex bound treo is neutralized by equal number of v^2 kinetons; which are present in v layered kinetic coloumn at each apex bound treo in Earth orbit.

i.e. 3.3999573×10^{78} free treos load is **supported by equal number of 3.3999573 × 10⁷⁸ Kinetons i.e. v² kinetons,** in one $1.84319077 \times 10^{39}$ layered kinetic coloumn of; i.e. **v number of layers.**

NOTE DOWN; v^2 *is* = MG/r, *according to Newton's equations,* and same is calculated here as kinetons in one kinetic coloumn which form at each apex bound treo in orbit of earth to neutralise the exerted load.

Therefore

1. **$1.84319077 \times 10^{39}$ quanta load is at quantum level of Sun where earth orbit is present.**

2. $1.84319077 \times 10^{39}$ or **v number of layers** have **3.3999573 × 10⁷⁸ Kinetons v^2** , in each kinetic coloumn present at each one apex bound treo in orbit of Earth.

8. At One RC Wave Length in Earth Orbit It Can Support Load of MG: Total Mass of Sun

In any n^{th} **layer** (from sun) **n number of kinetic coloumns** (each having v^2 kinetons) are present on each of n number of apex bound treo in its RC wave length, and kinetic energy in all kinetic coloumns jointly supports **total MG mass pressure of Sun.**

On n apex bound treos in RC wave length ($0.925863862 \times 10^{46}$ kinetic coloumns) in this n^{th} layer from Sun $\times 3.3999573 \times 10^{78}$ kinetons in each such kinetic coloumn = $3.145233887 \times 10^{124}$ free treos mass energy of Sun* or total MG mass of Sun in Earth orbit.

*[Sun is $0.91379754 \times 10^{38}$ unit mass body $\times 3.44247205 \times 10^{86}$ free treos per one unit mass = **$3.145233887 \times 10^{124}$ free treos** = 1.989×10^{30} kg is mass of sun $\times 1.5808523 \times 10^{94}$ free treos in one Kg; (as 1.5808523×10^{94} kinetons per Kg per sec per second which is derived conventional value of G)] (Refer to page 218)

9. On 2 RC Wave Length One Full Matter Wave Support Total 2 MG Exerted Load of Sun

Total (M^2) load at gravitational center of Sun, exerts 2MG load on any one concentric layer of gravitational (field) in graviton coloumn of sun and it is distributed as mass pressure of Sun, on 2n-1 apex bound treos of this layer, and this load is supported by 2n-1 kinetic coloumns which form one matter wave.

Thus **one matter wave** in $0.925863862 \times 10^{46}$ th layer of gravitational field on its $2 \times 0.925863862 \times 10^{46}$-1 apex bound treos (**at its 2 n–1 apex bound**) in one layer of gravitational (coloumn) field of Sun, supports total 2 MG mass pressure of Sun, exerted on earth orbit.

$2 \times 0.925863862 \times 10^{46}-1$ kinetic coloumns $\times 3.3999573 \times 10^{78}$ kinetons in each kinetic coloumn = $2 \times 3.145233887 \times 10^{124}$ **kinetons**; these kinetons support equal load of 2 MG free treos of Sun on Earth orbit.

10. Circumference of Earth Orbit

(A) **Crompton wave length of all the matter waves** at one matching quantum level in all graviton coloumns (which merge together to form gravitational field of Sun), **join side by side and together form full circumference of Earth orbit.**

Gravitational field of Sun is formed by union of $0.91379754 \times 10^{38}$ graviton coloumns (as Sun is made up of $0.91379754 \times 10^{38}$ unit masses).

In each of these **graviton coloumns** at 1.00658064×10^4 quantum level at a distance of 1.01320458×10^8 bound treo layer 'r' [r = $(1.00658064 \times 10^4)^2$] the Compton wave length **(2 π r) is (2 $\pi \times 1.01320458 \times 10^8$)**

In the **Sun's one graviton coloumn in its gravitational field** (formed by union of $0.91379754 \times 10^{38}$ graviton coloumns), at *distance of Earth ($0.925863862 \times 10^{46}$ bound treo layers =151.23 million Km) which is also the radius of earth orbit* [= $0.91379754 \times 10^{38} (1.01320458 \times 10^8)$ bound treo layers] the the Compton wave length **(2 π r) is** $2 \pi \times [0.91379754 \times 10^{38} (1.01320458 \times 10^8)$ bound treo layers] which form **circumference of orbit of earth = 5.819×10^{46} Bound treos.**

(B) **Earth orbit is Compton wave length** (2 × Reduced Compton wave length × π) **of matter wave in orbit of Earth = Circumference of Earth orbit.**

One matter wave in orbit of Earth forms on 2n–1 apex bound treos; and when mutiplied by pi; π (2n-1) = is Compton wave length of this matter wave in orbit of Earth = **2 × ($0.925863862 \times 10^{46}$ bound treos – 1) × 3.14 (value of π) = circumference of Earth orbit = 5.819×10^{46} Bound treos.**

(C) This Earth's orbit 5.816×10^{46} bound treos circumference, in which Earth revolves to complete its one revolution in one year = $1.843190775 \times 10^{39}$ bound treo distance per second × 31557600 seconds are in 365.25 days of one year.

For Understanding

(a) It is interesting to note, that **for the equal load placed at identical (same) quantum levels in two fields** (load of a elementary particle at one apex bound treo in matter wave in second dimension or mass pressure of Sun at one apex bound treo in matter wave which form in two dimensional deformation of gravitational field of Sun) the **wave which form in quantum world in two dimensional deformations, is also formed in two dimensional deformation of gravitational field of Sun** in which planets are placed at planetary orbits formed.

One small orbit of this wave in one graviton coloumn at 1.00658064×10^4 quantum level of Sun, **with the circumference (2 × 1.01320458 × 10^8 − 1) π ×** when **multiplied by 0.91379754×10^38** (number of all graviton coloumns of Sun which **unite side by side** to form gravitational field of Sun) = 5.816×10^{46} bound treos is **circumference of orbit of Earth**, on which Earth revolves in one year, by moving $1.84319077 \times 10^{39}$ bound treo distance per second ($\approx$ 30 Km per sec).

(b) r or n is distance of Earth's orbit from Sun, $0.925863862 \times 10^{46}$ bound treos layers.

But this **r or n, is also Reduced Compton wave length** of matter wave in this orbit **on which MG mass of Sun is supported.** (On twice of this RC wave length or on 2n-1 bound treos, one matter wave of Sun in Earth orbit is formed, and when multiplied by π it calculates circumference of Earth orbit).

We see same in quantum world, all elementary particle packet have their RC wave length n, which is calculated in bound treo distance (S number of bound treos/number of quanta in packet) and **on which total mass energy of elementary particle spreads. On twice of RC wave length, on 2n-1 bound treos one matter wave forms**, and when multiplied by π it calculates circumference of orbit.

These observation are, mathematical confirmation of this quantum gravitation model, and unites quantum physics with general theory of relativity.

11. Revolution Speed of Earth

(A) Orbital motion of electron

In motion of unit electron, in second dimensional deformation is by its matter wave.

Out of S vibrations which occur in one second, each shell present on each apex bound treos along its wave length vibrates one by one by **one supporting vibration per shell,** and thus by √S vibrations the 'wave packet of unit electron' rotates once.

After this one rotation, **by next propelling one vibration**, the unit electron packet is pushed to next bound treo in orbit, and this process is repeated, up to √S times in one second (in S vibrations). Thus, it pushes unit electron mass by √S bound treos distance in its orbit in one second. Unit electron rotates √S times and revolves in its orbit by √S bound treo distance in one second.

Frequency of matter wave in orbit = number of quanta load exerted at this quantum level = number of bound treo layers in each kinetic coloumn placed at each apex bound treo in its wave length = equal to the orbital speed when calculated in terms of 'number of bound treo distance per second'.

(B) Orbital motion of Planet Earth

As we have studied the orbital motion of unit electron, similarly the **symbolic point mass** (after neutralization of mass pressure of Earth on space matrix by its own gravitational sphere) of **Planet Earth, will remain stationary** at $1.00658064 \times 10^{4th}$ quantum level, till 1.00658064×10^4 vibrations occur one by one in each of 1.00658064×10^4 shells, which are present at each apex bound treo along its Reduced Compton wave length. And only after all these supporting vibrations by next one propelling vibration the point mass of Earth will shift to next one bound treo in its orbit on matter wave. (Fig.69)

This process will be repeated $1.84319077 \times 10^{39}$ times in one second (as total 1.85539×10^{43} vibrations occur in one second), and thus the **point mass of Earth will move in its orbit by $1.84319077 \times 10^{39}$ bound treo distance or 29.7901838* Km per second** which is the known revolving speed of earth.

$1.84319077 \times 10^{39}$ is the frequency of matter wave, $1.84319077 \times 10^{39}$ quanta is load of Sun at this quantum level, $1.84319077 \times 10^{39}$ are number of kineton layers in each kinetic coloumn in wave.

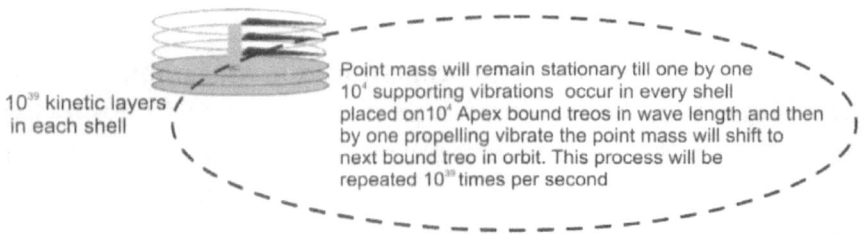

10^{39} kinetic layers in each shell

Point mass will remain stationary till one by one 10^4 supporting vibrations occur in every shell placed on 10^4 Apex bound treos in wave length and then by one propelling vibrate the point mass will shift to next bound treo in orbit. This process will be repeated 10^{39} times per second

10^{39} least lengths (30 Km) in 10^{43} vibrations = Orbitarthal Speed of Earth in 1Second

Figure 69: The 'point mass of planet Earth' shifts by 10^{39} bound treo distance in one second, equal to the frequency of matter wave in orbit.

*(Planck least length is 1.6162×10^{-35} meter; so, the total number of Bound treos present per meter $= 1/1.6162 \times 10^{-35} = 0.6187 \times 10^{35}$ bound treos per meter)

NOTE DOWN; Orbital speed of Earth, as according to Newton's equations is same as explained here according to treo model.

12. Biggest Planet of Our Solar System Broke Down

It can be noted that all four 'outer planets' i.e. Jupiter, Saturn, Uranus and Neptune have their biggest satellites at energy level ES1/9 or ES1 /3² where as this energy level Em/9 or Em/3² (where **Em is energy in orbit of planet Mercury** at each apex bound treo) in solar system harbors 'asteroid belts'.(**Where ES1 is energy in orbit of first sattelite at each apex bound treo of all four planets**)

The biggest satellite of all outer planets Jupiter – **Ganymede of 5262 Km** diameter, Saturn –**Titan, 5780 Km** diameter, Uranus –**Titania, 1578 Km** diameter, and Neptune–**Triton 2706 Km** diameter.

1. So, we can conclude that this quantum level at energy level **E/9 or (E /3²)** in any gravitational field normally harbors its biggest baby body.

2. With this logic, it can be presumed that our solar system also had its biggest Planet, **at Em/9 or (Em /3²)** energy level. But on this level now we found 'asteroid belt'. So, all asteroids may be the debris (pieces) of broken biggest planet[30] of our solar system, which would have occupied this energy level.

Uneven density of body of Moon can be attributed to this debris. It could also be inferred, that the *sudden change from single cellular life on earth to multi cellular life, in this early period* (possibly four billion years back) may have been **seeded on earth from this broken planet.**

---

30 This planet was Megaskar (as named by few astro-physicists), which broke down four billion years ago.

13. The Visible Rings of Saturn

Due to less density full mass of these cosmic bodies could not be contained in it's three dimensional spherical deformations, so the excess gaseous mass and debris of these bodies overflows and occupy the second dimensional deformation of their gravitational field and condenses along orbitums. Satellites also start condensing side by side in second dimensional deformation and are often seen inside the Saturn rings. Saturn form spectacular Rings along its orbitums. Though invisible from earth outer planets also form Rings from its own debris in all outer planets of Solar system.

14. Gravitational Coloumn of Outer Planet Tilts After Presence of One 'Heavy Satellite'

The gravitational coloumn of parent body tilts distally, after presence of one 'heavy baby satellite' in its gravitational (field) coloumn. This can be verified, by tilted inclination angel of orbits of satellites, immediately after presence of a heavy satellite in gravitational field of all 'outer planet'[31].

This Is Visible Proof Of Formation Of Gravitational Coloumn, And Their Tilting In Second Dimensional Deformation Of Gravitational Field Of Any Body. (Fig.70)

---

31 Refer to table of satellites in my previous book 'How universe works – quantum gravitation and fifth dimension' Manes Prakashan,1915. to confirm the 'tilted inclination angle of 'orbits of distal satellites of outer planets' immediately after presence of a heavy satellite.

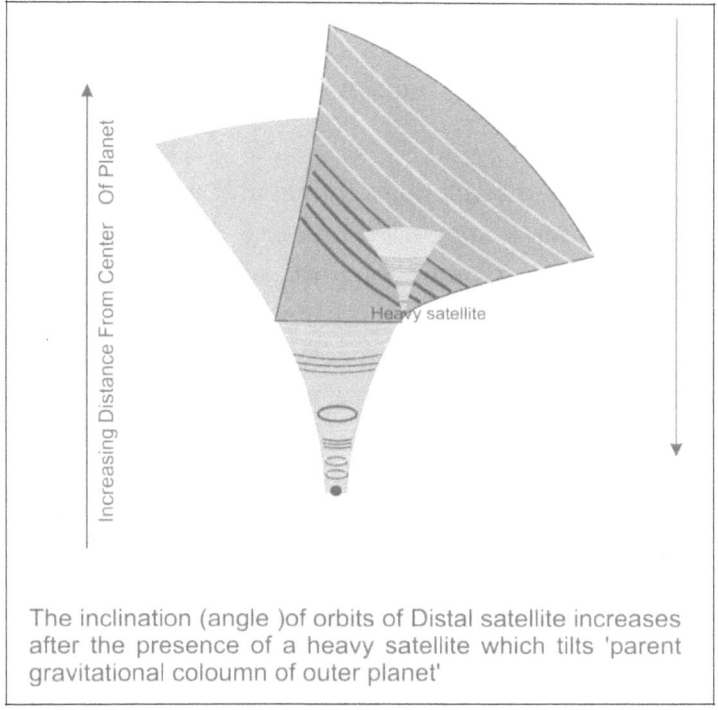

The inclination (angle)of orbits of Distal satellite increases after the presence of a heavy satellite which tilts 'parent gravitational coloumn of outer planet'

Figure 70: A heavy satellite will tilt the parent gravitational coloumn of planet.

15. Our Galaxy & Its Rotational Velocity Curve: A 88 Year's Old Unsolved Puzzle

The present understanding of mass/energy of universe is clarified in treo model by ten dimensional space matrix and free treos (11th dimension?).

Broadly these are responsible for (!) Hubble constant, cosmological constant and dark energy (**voids of treo model**) (!!) dark matter (27% mass energy of universe; with a average density of one proton mass in one sq. meter); it includes all bound treos of space matrix and its all formations; i.e. sub kinetic coloumns, Magnetic fields & magnetrons, all bosons as orbitums and orbits, gluons and W, Z particles, C mue, Higgs boson, gravitons, electron black hole, gravitational spheres, black holes and *dark halos*, cosmic strings and filaments, as all are from **Bound treos of space matrix.** (!!!) Fermions constituting ordinary matter and anti matter, electrons, charge, nucleons, atoms and unit masses, all are formed from **free treos.**

The Galaxy made up of black holes (any n unit masses) and its load in square of unit masses is supported by equal dark mass in *dark halos* (as solar gravitational coloumn supports our Sun); by all **gravitons.** Our Sun is placed in our spiral galaxy, milky way (100,000 light years across: second largest after Andromeda in local group), at 27000 light years away from galactic center in a minor arm and in a small spur Orion spur (with four main spiral arms of Milky way i.e. Norma and Cygnus, Sagittarius, Scutum-crux and Perseus).

We could not answer the riddle of galactic velocity curve (Rotation curve of galaxy, of increasing orbital speed of stars, with their increasing distance from galactic centre) **by our present knowledge of gravitation.** Newton described the gravitational field equations ($MG = r^2 a$; $MG = rv^2$) and these equations are valid for study of one dimensional gravitational field in Gravitational coloumn of Sun. Einstein gravitation is description of this field in curved space time, with its causality at centre of this curved space.

But gravitation is four dimensional deformation of space matrix to carve self supporting universe, its all galaxies, black holes, Sun, planets on space matrix. (as described above in Treo model).

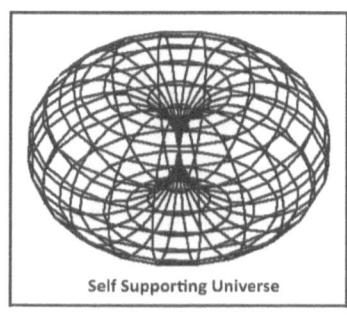

Self Supporting Universe

Figure 71: Self supporting universe is Architectural marvel of space matrix

The **Keplerian dynamic** is valid only for planets and satellites and on stars (e.g. of Star Trappist 1; described in detail in appendix 1) in which orbital velocity of planets 'v' ($v = \sqrt{v^2} = \sqrt{MG/r}$) is decided by gravitational kinetic energy or v^2. But in a galaxy the stars are placed in three dimensional deformations (dilution) of galactic gravitational (coloumns) field.

Calculation of gravitational kinetic energy in solar gravitational field (formed by orbits, shells and transverse sub kinetic coloumn) are done by Newton's equation ($MG = rv^2$); while the cyclonic wave of its third dimensional spherical deformation harbors' body of Sun.

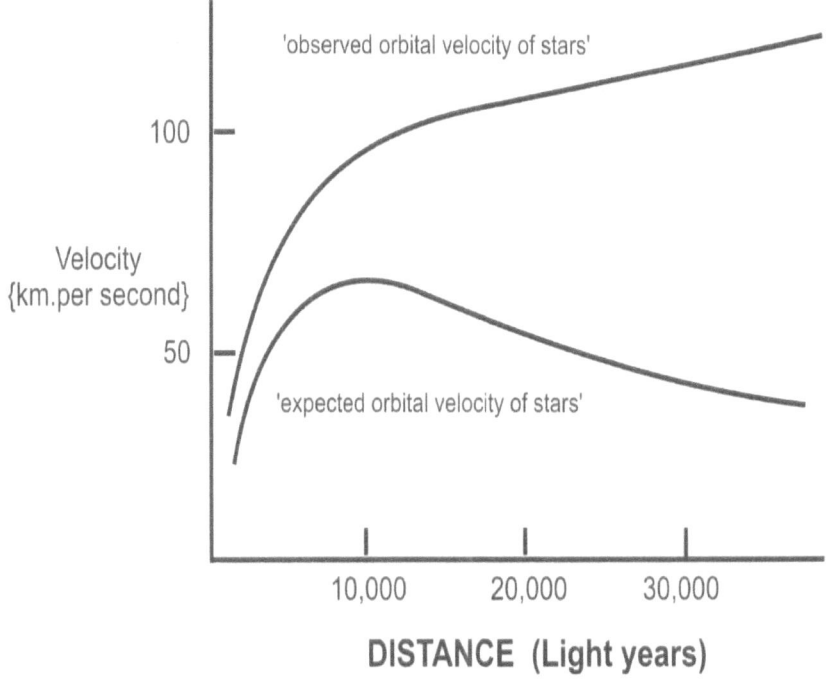

Figure 72: Rotational curve of galaxies

[Velocity of star does not decelerates with distance as expected, according to solar gravitational field ($v = \sqrt{MG/r}$); but it is now explained by four dimensional coloumn geometry of treo model which form Galactic gravitational coloumn.]

Cyclonic waves of all black holes (in third dimensional deformation) which jointly support galaxy, are formed in rotating spiral kinetic coloumn of third dimension (electron black hole) with one graviton at its each bound treo in gravitational field. Each kinetic coloumn directs its total force of all gravitons towards its gravitational centre; so keplian kinetics its rules and formula ($MG = rv^2$) will not work in three dimensional deformation of gravitational field of galaxy.

In first and second dimensional deformation of gravitational fields the gravitational kinetic energy v^2 fades with distance by 'inverse square law', due to **gradual reduction of layers of kinetic coloumns** in peripheral fields; but with the addition of $2n-1$ gravitons in any n^{th} layer at n^{th} quantum level in the three dimensional deformation of gravitational field the ground energy due to gravitons (S^2 kinetons) at each bound treo will remain **same,** but their number will increase in each peripheral layer. Thus the rotating gravitational coloumn of galaxy (dark halo) will behave as solid sphere (like three dimensional body of solar sphere), in which embedded stars rotate. In such sphere the different peripheral layers rotate at increasing higher speeds and so the velocity of stars embedded in periphery will be higher.

Now this combined kinetic energy in these three dimensional gravitational fields (which make spiral arms of galaxy) over rules the problem of reducing orbital velocity of stars with its increasing distance.

As explained in detail, in my previous book ('Our universe and how it works: Quantum gravitation and fifth dimension', Manas Prakashan, India, 2015, ISBN 978-9-35235-003-2 and e. book, book baby, 2016, ISBN 978-9-35235-065-0,) **inter galactic journeys will be possible** (for our space ships made up of polymers and assembled on moon) via empty spaces in between spiral arms, which will serve as our galactic high ways (very easy approachable from solar system as Sun is situated in a spur placed by the side of this highway).

For any query regarding this research work, please contact:
mail2ashoksaxena@gmail.com
91-9837153974

Contraction of Unit Space Matrix	Unit of Increment of mass Energy	Particle generated and supported	Wavelength or Spread of Mass pressure	Load exerted on each apex bound treo along wavelength	Ground energy in the dimension	Geometry of deformation	Fields produced
First dimensional deformation of LENGTH	S free treos, which is same as one quanta mass energy	All photons of EM spectra	Decreasing Reduced Compton wavelength: From S bound treos, Till √S bound treos	From one free treo on each bound treo, Till S free treos load on each apex bound treo along wavelength of photon packet	One kineton, which is same as one deformed bound treo	a. Sub kinetic Coloumns, b. Layers of kinetic coloumn and c. Kinetons (which form transverse EM Waves)	EM Fields

	One Unit	Unit	Decreasing	From one	S Kinetons		Atomic & Weak Forces
Second dimensional deformation of LENGTH & BREADTH	Electron ($\sqrt{S}$ × S Free treos)	Electron, All elementary particles, Nucleons, All Elements & Molecules, up to One– unit mass.	Reduced Compton wavelength: From $\sqrt{S}$ bound treos, Till one bound treo (one graviton	quantum on each apex bound treo, Till S quanta on each apex bound treo along wavelength of elementary particle	in one 'orbitum'	a. Shells, b. Sub–shell and c. Orbitums (together form Matter waves)	Forces
Third dimensional deformation of LENGTH, DEPTH & BREADTH	One Unit Mass (S × S Free treos)	From one Unit Mass, which is Planck mass, Till $\sqrt{S}$ Unit masses	ncreasing wavelength: From one bound treo, Till $\sqrt{S}$–1 bound treos	From one–unit mass or S Quanta, Till S unit masses at gravitational center.	S² Kinetons at one graviton	a. Electron Black Hole, b. Spiral circular layers of electron black Holes, c. Gravitons	Start of formation of Gravitational Fields

Continued

Deformation of Four dimensions of SPACE–TIME	One Electron Black Hole ($\sqrt{S}$ × S × S free treos)	From $\sqrt{S}$ unit masses, Till S unit masses in one Unit black hole	Increasing wavelength: From $2\sqrt{S}-1$ Bound treos, Till $2S-1$ bound treos in last layer	From S unit masses, to S^2 unit masses at gravitational center	S^3 kinetons on one electron black hole	a. Gravitational spheres, (largest being one–unit black hole) b. Layers made of electron black holes, c. Electron black holes (each having S gravitons)	Gravitational Spheres

Appendix 1

Study of Gravitation Forces Outside Our Solar System (Study of Gravitational Field of Star Trappist 1)

As we observe, that any mass 'm' exert a load of m^2 mass pressure on space matrix at it gravitational centre, which is supported by formation of a reacting gravitational sphere from space matrix made up of equal number of of (m^2) gravitons. This central load dilutes as 2 MG mass pressure on its each layer of gravitational field, where it spreads equally on each of 2n-1 apex bound treos in layer exerts a **load r a** on each apex bound treo at distance r from body. it is **supported by equal number of kinetones v^2** in each reacting sub kinetic coloumns which neutralise this load.

After study of positioning pattern of all planets in 2 dimensional deformation of gravitational field in our solar system and also verifying this pattern in gravitational fields of all 'four outer planets' (by studying positions of their satellites) [ref 6], we found the same pattern when **we move outside of our solar system.**

The Earth like seven planets, were discovered by Gillon et al on 22 feb 2017, with the help of Spitzer space telescope, which were orbiting around an ultra-cool red dwarf star named as Trapist 1; present at 39.6 light years away in constellation Aquarius.

Analysis of Data of seven planets of Trapist 1 (1b, 1c,1d,1e,1f,1g,1h), will be done according to treo model, to confirm the proposed quantum model of gravitation, outside our solar system.

(A) Mass of Star Trappist 1 in Kg

According to given data, *the mass of star is 0.0802 (+ – 0.0073) of that of Sun or 84 times mass of Jupiter,* {Ref 8}

(a1) Mass of Trappist 1 is 0.0802 then the mass of our Sun = Mass of Sun in Kg × 0.0802 = 1.989×10^{30} Kg × 0.0802 = **159.51×10^{27} Kg**

(a2) Mass of Trappist 1 is approximately = Mass of Jupiter in Kg × 84 times = $1.8986 \times 10^{27} \times 84 =$ **159.4824×10^{27} Kg**

The 159.51×10^{27} Kg mass of star is contained in spherical three-dimensional deformation of gravitational field of the star, which forms **spherical body of star.**

(B) Mass of star Trappist 1 in terms of free treos

According to proposed model one Kg mass is made up of 1.580852×10^{94} energy particles (Free Treos) [Please see page 48]

Number of energy particles forming mass energy of Trappist 1 = 159.51 × 1027 Kg × 1.580852×10^{94} energy particles in one KG = **$252.161750 \times 10^{121}$** energy particles (Free treos) in Trappist 1

(C) Mass of star Trappist 1 in term of unit masses

According to treo model One unit mass have S^2 or $3.4405427169 \times 10^{86}$ energy particles as free treos.

$252.161750 \times 10^{121}$ is number of free treos as mass energy in star $/3.4405427169 \times 10^{86}$ free treos in one unit mass. = **73.291272×10^{35} unit masses** in Trappist1Star .

(D) Distance of all 7 baby bodies in bound treo layers.

[according to provided data of distance in AU 10^{-3} ; 11.11, 15.21, 21.44, 28.17, 37.1, 45.1, 63 (+ – 13 to 26)]

1 AU = Distance of Sun from Earth = $0.9257897548 \times 10^{46}$ bound treo layers of Space matrix from Sun, in Gravitational field (coloumn) of Sun.

(1b) Distance of first planet of Tappist 1b calculated in terms of bound treo layers = 11.11×10^{-3} AU = $11.11 \times 10^{-3} \times 0.9257897548 \times 10^{46}$ =

1.0285524×10^{44} bound treo layers from star Trappist 1 Distance of other planets of Trappist – 1 (1c,1d,1e,1f,1g,1h)

(1c) 15.21×10^{-3} AU $\times 0.9257897548 \times 10^{46}$ = **1.408126×10^{44} Bound treo layers**

(1d) 21.44×10^{-3} AU $\times 0.9257897548 \times 10^{46}$ = **1.984893×10^{44} Bound treo layers**

(1e) 28.17×10^{-3} AU $\times 0.9257897548 \times 10^{46}$ = **2.607949×10^{44} Bound treo layers**

(1f) 37.10×10^{-3} AU $\times 0.9257897548 \times 10^{46}$ = **3.434679×10^{44} Bound treo layers**

(1g) 45.01×10^{-3} AU $\times 0.9257897548 \times 10^{46}$ = **4.166979×10^{44} Bound treo layers**

(1h) 63×10^{-3} AU $\times 0.9257897548 \times 10^{46}$ = **5.832475×10^{44} Bound treo layers**

(E) Number of bound treo layers in four-dimensional deformation which form gravitational sphere of star Trappist 1

According to treo model, the gravitational sphere of any 'n' unit masses cosmic body have 'n' number of bound treo layers.

Number of bound treo layers in gravitational sphere of star Trappist 1 ;

73.291272×10^{35} unit masses are in Trappist 1 star = thus $73.291272 \times$

10^{35} bound treo layers ≈ (120 meter Radius) form gravitational sphere of this star Trapist 1

(F) Position of baby bodies

In two-dimensional deformation of gravitational field of star Tappist 1

According to treo model First baby body (Star, Planet or Satellite) of any cosmic body (galaxy, star or Planet) condenses at 10^{4th} quantum level of gravitational field of parent body.

This distance from parent body to first baby body can be calculated in bound treo layers by formula (number of bound treo layers in gravitational sphere of Star × square of 10^{4th} quantum level)

In Support of the treo model, *similar to first planet of Sun i.e. Mercury (and also similar to presence of First Satellite of all four outer planets) the*

First planet of Trapist 1 is also condenses at 0.3746166 × 10^4 quantum level, and its distance from parent star can be calculated according to proposed formula.

Distance of First planet of Trappist 1 from parent star = 73.29127247611 × 10^{35} bound treo layers are in its gravitational sphere × (0.3746166 × 10^4)^2 = 1.028552 × 10^{44th} bound treo layer from centre of Trappist 1.

According to treo model the gravitational sphere of any 'n unit mass cosmic body' have n layers in its gravitational sphere, which together supports its total n^2 unit mass load at its gravitational centre.

The gravitational (coloumn) field of any cosmic body is formed by union of 2n-1 graviton coloumns of 2n-1 gravitons (each supporting one unit mass) present at periphery of n layered gravitational sphere of this n unit mass cosmic body.

Compton wave length of all matter waves, in these 2 × (73 × 10^{44})-1 graviton coloumns together unite to form orbit of planet, placed at this planetary quantum level.

Or

The circumference of this planetary orbit (Reduced Compton wave length of matter wave of this planetary orbit × 2 π) as we can omit − 1 from equation being very small.

(G) Circumference of ORBITS.

(RC wave length of matter wave of this planetary orbit × 2 π;

Or Distance of planet from star in bound treo layers or radius of its gravitational coloumn upto particular planet × 2 π.)

(Circumference of orbit or Compton wave length of orbit of first planet1b) = 1.028552 × 10^{44} Bound treos is RC wave length of matter wave in orbit × 2 π = 6.465184 × 10^{44} bound treos.

Similarly −

(1c) 1.408126 × 10^{44} RC wave length ×2 π = **8.851079 × 10^{44} bound treos**

(1d) 1.984893 × 10^{44} RC wave length × 2π = **12.476471 × 10^{44} bound treos**

(1e) 2.6079495 × 10^{44} RC wave length × 2π = **16.392826 × 10^{44} bound treos**

(1f) 3.4346795 × 10^{44} RC wave length × 2π = **21.589416 × 10^{44} bound treos**

(1g) 4.1669795 × 10^{44} RC wave length × 2π = **26.192443 × 10^{44} bound treos**

(1h) 5.832475 × 10^{44} RC wave length × 2π = **36.661272 × 10^{44} bound treos**

(H) Gravitational kinetic energy in orbits.

Number of kinetons (total Kinetic energy) in any one sub kinetic coloumn, which are present at each apex bound treo in matter wave of orbit, is calculated as v^2 (where v^2 = MG/r)

[We will study the gravitational field of Star Trappist 1 with the example of calculations in orbit of first planet MERCURY IN SOLAR SYSTEM as per treo model –

Kinetic energy in each sub kinetic coloumn in the orbit of Mercury is calculated = Em = v^2 (according to Newton's formula $v^2 = MG/r$) = MG of Sun is $3.145233887 \times 10^{124}$ free treos/$0.3583722239 \times 10^{46th}$ bound treo layers i.e. r distance of Mercury from Sun = $8.776444393 \times 10^{78}$ kinetons or v^2 kinetons are in each v layered sub kinetic coloumn which is present at each apex bound treo and form one matter wave in orbit of Mercury.

2MG mass pressure of Sun is exerted in any one direction, which gets equally distributed at 2n-1 apex bound treos, and is supported by total kinetic energy (by v^2 kinetons from each 2n-1 kinetic coloumns), at distance r (or n bound treo layers from sun) by common matter wave present in the orbit of mercury.

This kinetic energy v^2 in each kinetic coloumn is generated by space matrix in response to load of body (r × a) at this apex bound treo ('r' or distance from gravitational center × 'a' or *diluted mass pressure of body at this distance.)

***This explains action – reaction mechanism of space matrix:** when r × a is action of load on space matrix at this apex bound treo = v^2 is reaction of local space matrix at this point by which form one sub kinetic coloumn.*

*This calculation can also be done by Newton's formula equations ($MG = r$ v^2 and $MG = r^2 a$ i.e. $ra = v^2$) which also explains action reaction mechanism by this equation **ra** $=v^2$, at any point in any gravitational field.*

*(in equation, a = MG/r^2. 'a' is Newton's acceleration, or Einstein's Slope of deformation, or 'diluted mass pressure of central load of body at this distance according to treo model.)

In the orbit of Mercury in gravitational field of Sun this action reaction mechanism can be visualized, ra = v^2

('r' or 0.3583722239 × 10⁴⁶ × a' or 24.31965189 × 10³² free treos = 8.776444393 × 10⁷⁸ free treos) = (8.776444393 × 10⁷⁸ kinetons or v² kinetons)]

Similarly v² is kinetic energy in each sub kinetic coloumn of matter wave in orbit of first planet 1b of Trappist 1= MG/r

(1b) $252.161750 \times 10^{121}$ free treos is mass of star in free treos i.e.

$MG/1.028552 \times 10^{44}$ (or r) = 24.5161×10^{78} kinetons.

Calculation of **V²** (i.e. gravitational kinetic energy present in each sub kinetic coloumn of matter wave in orbit) at other planets of Trapist 1

(1c) $252.161750 \times 10^{121}$ free treos/1.408126×10^{44} (or r) = 17.907609×10^{78} kinetons (v² kinetons)

(1d) $252.161750 \times 10^{121}$ free treos/1.984893×10^{44} (or r) =12.704046×10^{78} kinetons (v² kinetons)

(1e) $252.161750 \times 10^{121}$ free treos/2.607949×10^{44} (or r) = 9.668965×10^{78} kinetons (v² kinetons)

(1f) $252.161750 \times 10^{121}$ free treos/3.434679×10^{44} (or r) = 7.341637×10^{78} kinetons (v² kinetons)

(1g) $252.1617503 \times 10^{121}$ free treos/$4.16697968 \times 10^{44}$ (or r) = $6.051427406361 \times 10^{78}$ kinetons (are v² kinetons)

(1h) $252.1617503 \times 10^{121}$ free treos/$5.83247545 \times 10^{44}$ (or r) = $4.22340868872 \times 10^{78}$ kinetons (are v² kinetons)

(I) **Orbital speed of planets**

ACCORDING TO TREO MODEL Frequency of matter wave at any n^{th} quantum level = Number of quanta load at each apex bound treo = number of bound treo layers in any one sub kinetic coloumn = **Orbital speed (Number of bound treos per second) of planet.**

Frequency $=\sqrt{v^2}$ = v = frequency of matter wave = **orbital speed of planet in orbit** (is calculated both in bound treo distance per sec. and in Km per sec)

(1b) Orbital speed of first planet of Trappist 1 in bound treo distance per sec and in Km per sec

$(24.5161 \times 10^{78}$ kinetons$)^{0.5}$ = v = 4.951373×10^{39} bound treo layers in each sub kinetic coloumns/$0.618724203 \times 10^{38}$ bound treos in one Km length = **80.00255253 Km per sec**

Orbital speed of all other six planets in bound treo distance and also in Km

(1c) $(17.907609 \times 10^{78}$ kinetons$)^{0.5}$ = v = 4.231738×10^{39} bound treo distance per sec/$0.618724203 \times 10^{38}$ bound treos in one Km length =

68.3945768 Km per sec

(1d) $(12.704046 \times 10^{78}$ kinetons$)^{0.5}$ = v = 3.564273×10^{39} bound treo distance per sec/$0.618724203 \times 10^{38}$ bound treos in one Km length =

57. 6068139 Km per sec

(1e) $(9.668965 \times 10^{78}$ kinetons$)^{0.5}$ = v = 3.109495×10^{39} bound treo distance per sec/$0.618724203 \times 10^{38}$ bound treos in one Km length =

50.256559 Km per sec

(1f) $(7.341637 \times 10^{78}$ kinetons$)^{0.5}$ = v = 2.709545×10^{39} bound treo distance per sec/$0.618724203 \times 10^{38}$ bound treos distance in one Km =

43.792452 Km per sec

(1g) $(6.051427 \times 10^{78}$ kinetons$)^{0.5}$ = v = 2.459964×10^{39} bound treo distance per sec/$0.618724203 \times 10^{38}$ bound treos distance in one Km =

39.758651 Km per sec

(1h) $(4.223408 \times 10^{78}$ kinetons$)^{0.5}$ = v = 2.055093×10^{39} bound treo distance per sec/$0.618724203 \times 10^{38}$ bound treos distance in one Km =

33.21500 Km per sec.

(J) Time required by planet to complete its one orbit,

(revolution time of each plant as calculated in seconds and in Earth's day)

(Circumference of orbit in bound treo distance/speed of planet in bound treo distance per sec)

and

(also in one Earth's day, where one Earth day = 60 sec × 60 minutes × 24 hour = 86400 sec)

(1b) 6.465184×10^{44} bound treos is circumference of orbit/ 4.951373×10^{39} bound treo distance per sec = 1.305735×10^{5} Sec/86400 sec in a day = **calculated/ Actual 1.51 earth day.**

(1c) 8.851079×10^{44} bound treos is circumference of orbit/ 4.231738×10^{39} bound treo distance per sec = 2.09159408855 Sec/86400 sec in a day = **calculated/ Actual 2.42 day.**

(1d) 12.476471×10^{44} bound treos is circumference of orbit/ 3.564273×10^{39} bound treo distance per sec = 3.500428×10^{5} Sec/86400 sec in a day = **(calculated 4.04 day) Actual 4.05141day.**

(1e) 16.392826×10^{44} bound treos is circumference of orbit/ 3.109495×10^{39} bound treo distance per sec = 5.271859×105 Sec/86400 sec in a day = **(calculated 6.06 day) Actual 6.101689 day.**

(1f) 21.589416×10^{44} bound treos is circumference of orbit/ 2.709545×10^{39} bound treo distance per sec = 7.96791×10^{5} Sec/86400 sec in a day = **(calculated 9.1 day) Actual 9.222 day**

(1g) 26.192443×10^{44} bound treos is circumference of orbit/ 2.459964×10^{39} bound treo distance per sec = 10.647486×10^{5} Sec/86400 sec in a day = **(calculated 12.35 day) Actual 12.32 day**

(1h) 36.661272×10^{44} bound treos is circumference of orbit/ 2.055093×10^{39} bound treo distance per sec = 17.83922×10^5 Sec/86400 sec in a day = **(calculated 20.62 day) Actual 20.647 day**

(The discrepancy in calculated values and observed values is due to gravitational effect of other planets on orbital speed of observed planet.)

(K) Kinetic energy ratio of planets

Planets of Trappist 1 are placed in *s, p, d, f* sub shells according to their calculated kinetic energy ratio with first planet $v^2 1b$.

(1b) 24.5161×10^{78} kinetons (or $v^2 1b$ kinetons is gravitational kinetic energy of first planet 1b in its orbit) = *s1*

Ratio of Kinetic energy in first planet compared with kinetic energy in orbits of other planets gives a ratio $V^2 1b/n^2$ at n^{th} planetary quantum level

(This ratio of $V^2 1b/v^2$ kinetic energy in orbit of other planets of Trappist 1 is same as conventional value of energy distribution levels in *s,p,d,f.* atomic orbits).

(1c) 24.5161×10^{78} kinetons/17.907609×10^{78} kinetons = **1.369** = s1

(1d) 24.5161×10^{78} kinetons/12.704046×10^{78} kinetons = **1.929** = p1

(1e) 24.5161×10^{78} kinetons/9.668965×10^{78} kinetons = **2.535** = p2

(1f) 24.5161×10^{78} kinetons/7.341637×10^{78} kinetons = **3.339** = p3

Second quantum level

(1g) 24.5161×10^{78} kinetons/6.051427×10^{78} kinetons = ¼ of s1=

4.051292 = *s2*

(1h) 24.5161×10^{78} kinetons/4.223408×10^{78} kinetons = ¼ of s1 =

5.804813 = *s2'*

Appendix 2

Observations of Ancient Indians

There is a lot of evidence that ancient Indian civilization was using extremely accurate astronomical, heliocentric calculations, for both Earth and celestial motions, indicating an understanding that the **Sun is at the center of the solar system and that the Earth is round.**

As per the Vedas and ancient texts, it appears that ancient Indians have calculated accurately the motion of planets, sunset, sunrise, eclipses, etc. presumably without using telescopes or any other machinery. Elliptical orbits were also calculated for all moving celestial bodies. What Indians calculated thousands of years ago, for example the **wobble of the Earth's axis**, which creates the movement called **'precession of the equinoxes'** – the slowly changing motion that completes one cycle every 25,920 years – has only recently been validated by modern science.

When we talk of **gravity,** Newton comes to our mind, but in the text **Surya Sidhantha** dated around 400 AD, **Bhaskaracharya** described **"objects fall on the Earth due to one force. The Earth, planets, constellations, moon and Sun are held in orbit because of that one force".**

Forty–four centuries before Isaac Newton, the ancient Hindu text **Rig Veda written in 1750 BC (bronze age INDIA)** asserted that a force (which we now call gravitation) held the universe together.

Bhaskara I (629 AD): "Yavantamakasapradesam ravermayukhah samantat dyotayanti tavan pradesah khagolasya paridhih khakaksya anyatha hyaparimitatvat akasasya parimanakhya namnopapadyate".

It can be roughly translated as – That much of the sky as the Sun's rays illuminate on all sides is called the orbit of the sky. Otherwise, the sky is beyond limit; it is impossible to state its measure.

This implies that while the universe is infinite, but the solar system extends as far as the rays of the Sun can reach. (This can be the 'acceleration zone of the Sun according to present model' or the limit up to which the 'diluted mass pressure of Sun' and it's reacting 'gravitational (field) coloumn of Sun' finally extends, causing the deformation in first dimension.)

Spherical Earth: The Sanskrit speaking Aryans subscribed to the idea of a spherical Earth in an era when the Greeks believed in a flat one. The existence of rather advanced concepts, like the **spherical Earth and the cause of seasons were quite clear in Vedic literature.**

Aitareya Brahmana (verse 3.44) declares: "**The Sun does never set nor rise.** When people think the Sun is setting (it is not so). For after having arrived at the end of the day it makes itself produce two opposite effects, **making night to what is below** and **day to what is on the other side**. Having reached the end of the night, it makes itself produce two opposite effects, **making day to what is below** and **night to what is on the other side.** In fact, the Sun never sets."

(A) Seven colors compromising light: "**Seven horses draw the chariot of Surya**". **Rig Veda 5. 45.9:** These seven horses are the seven colors comprising light. These seven colors become visible in a rainbow or when light passes through a prism.

The velocity of Light was calculated by Maxwell in the 19th century, but it was actually determined accurately thousands of years before as mentioned in the **Rig Veda, and calculated accordingly as below.**

It was further elaborated **by Sayana in the 14th century AD in his commentaries on Rig Veda.** (The Rig Veda is the oldest Indian text and one of the oldest surviving in the world.)

Sayana (c.1315–1387) a vedic scholar and prime minister in the court of Bukka I (and his successors) of Vijaynagar empire in Karnataka of south India in his commentary says,

Fourth verse of the hymn 1.50 of Rig veda: 'Tatha ca smaryate yojananam. Sahasre dve dve sate dve ca yojane ekena nimisardhena kramaman'.

It means "it is remembered here that Sun (light) traverses 2,202 yojanas in half a nimisha" {Same statement occurs in the commentary of Bhatta Bhaskara (10th century) where it is said as old Puranic tradition.}

The **'yojana'** is an ancient unit of length. 'Arthshastra' defines it as equal to 8,000 dhanus, which is equivalent of 9.09 miles.

A **'nimisha'** is an ancient unit of Time, that is equal to 16/75 Sec.

Thus **'2,202 yojana in half a nimisha' is equal to 189,547 miles per second** (accepted speed of light today is 186,281.7 miles per second).

CALCULATIONS

In 'Moksha Dharma Parva' of Shanti Parva in 'Epic Mahabharata' (Nimisha, time taken for blink of eye)

15 Nimisha		= 1 Kastha
30 Kastha		= 1 Kala
30.3 Kala		= 1 Muhurta
30	Muhurta	= 1 Diva–Ratri (i.e. 24 hours)
15	×30×30.3×30	= 409050 Nimisha in 24 hours
While 60 × 60 × 24		= 86,400 Seconds in 24 hours
Thus 409050 Nimisha		= 86,400 seconds

Or one Nimisha is 0.2112 seconds and **half Nimisha is 0.1056 seconds.**

The **'Vishnu Purana' book 1 chapter 6 defines 'Yojana', a Vedic unit of distance**

10 Paramanus		= 1 Para–súkshma
10 Para súkshmas		= 1 Trasarenu
10	Trasarenus	= 1 Mahírajas (particle of dust)
10	Mahírajas	= 1 Bálá–gra (hair's point)
10 Bálá–gra		= 1 Likhsha
10 Likhsha		= 1 Yuka
10	Yukas	= 1 Yavodara (heart of barley)
10	Yavodaras	= 1 Yava (barley grain of middle size)
10 Yava		= 1 Angula (1.89 cm or approx 3/4 inch)
6	Angula	= 1 Pada (the breadth of foot)
2	Padas	= 1 Vitasti (span of fingers)
2	Vitasti	= 1 Hasta (cubit)
4	Hastas	= a Dhanu, a Danda, or Paurusa (a man's height)
or 2 Nárikás		= 6 feet
2,000 Dhanus		= 1 Gavyuti (distance to which a cow's lowing can be heard)
		= 12,000 feet

4 Gavyutis = 1 Yojana = 9.09 miles

Thus, speed of light

2,202 × 9.09 = 20,016.18 miles per 0.1056 seconds,

i.e.**189, 547 miles per second**

The **Indians of the fifth century A.D. calculated the age of the Earth as 4.3 billion years.** This culture gave us **zero,** the numerals that we use. So called Arabic have their roots in India – as do **trigonometry**

and calculus, astronomical calculation and a view that says the universe is not only billions, but trillions of years in age and that we are eternal beings who are simply visiting the material world to have the experience of being here.

Glossary

TREO – Primordial particle of creation, it is one dimensional of Planck's least length and constantly vibrates at Planck frequency; as **free treos** it forms all matter and as **bound treos** it represents five positive dimensions as it weaves 10 dimensional space matrix, while alternately arranged with voids.

VOIDS – Voids are alternately bound with treos to construct 10 dimensional space matrix, where it represent five negative dimensions of universe. They were fully curled at big bang and are constantly uncurling to expand our universe and it being dark energy assigns value of Hubble constant and cosmological constant.

Space Matrix – Omnipresent, omnipotent, ten dimensional matrix of universe which have three inter dependent components of SPACE–TIME–ENERGY.

Coloumn – One, two and three dimensional unit Space with fourth dimension of Space–time, by its gradual contraction adds one layer in the coloumn, which enlarges at $\sqrt{S}$ quantum levels, in each of four dimensions to form fields of all four forces, with $2n-1$ units in its each layer and n^2 units in one full **Kinetic coloumn**.

Kineton – One contracted bound treos changes to one kineton, when instead of vibrating in S planes it now only vibrates S times in the direction of exerted load to support it, by S vibrations for one second.

Orbital – Each orbital is made up of by S Kinetons) and supports load of one Quanta (S number of free treos) mass energy by its one rotation in one second, such 2n-1 units are in each sub shell (layer) of this two dimensional configration. All equal energy orbitums present one below other in all shells along its RC wave length, forms one orbit.

Graviton coloumn – One full graviton coloumn forms layer by layer at √S quantum levels and after its full formation in second (two) dimension at unit gravitational centre (now named as **one Graviton**) of **unit mass** (Planck's mass)

Unit photon – Packet of S number of free treos (**one quanta mass energy**)

Unit electron – Packet of √S × S number of free treos (**√S quanta mass energy**)

Unit mass – Packet of S × S number of free treos (**S quanta mass energy**)

Electron black hole (of the size of one electron √S bound treo layers) – Packet of S number of gravitons in √S layered kinetic coloumn of third (three dimensional deformation) dimension (i.e. **S quanta kinetic energy**) supports its One Billion ton mass energy i.e. √S quanta mass energy (√S × S × S free treos).

Unit black hole (biggest gravitational sphere i.e. kinetic coloumn of deformation in four dimensions of space –time; in 3 lac km radius) – Packet of S number of electron black hole or S × S number of gravitons, to support and churn a body of **S quanta mass energy packet (**S × S number of free treos).

How you define GOD, omnipresent, omnipotent, generator of universe, its operator and also the destroyer of every thing (in Indian mythology these powers are assigned to its espcific god known as Brahma, Vishnu and Shiva, respectively) . Are not space matrix have all these properties which are incoporated with treo, kinetons and voids ? Is Not space matrix is god of universe ?

SPACE – TIME-ENERGY

This trio vibrating at COMSIC RHYTHEM or S times per second, regulates the structure and working of Universe.

In all possible definitions, this TRIO represents three gods, as it generates all space, Time, all matter, all forces, all motion, it operates all mechanisms and even destroyes every thing.

But there is purpose behind every action and this puroseful actions in universe is proof of GOD.

BRAHMA – VISHNU-SHIVA (Treos, Kineton,Void) .... Generator, Operator, Destroyer

Brahma – as Treos – Generates of everything, all photons, matter, space–time–energy (space matrix).

Vishnu as – Kinetons – Operator – responsible for working of universe.

Shiva as – Voids – Destroyer, curled up five negative dimensions of universe which are slowly uncurling with each vibration of universe. It represents five negative dimensions as it has **NO length, NO breadth, NO depth** and it **calibrates gone time** of universe (as denoted by percentage of uncurling of each void of space matrix with passage of time), and it represents itself a negative **dark energy**. It is **responsible for aging of all living and non living things including of universe itself.**

References and Bibliography

1. *Roger Ariew; Ockham's Razor,* **A Historical and Philosophical Analysis of Ockham's Principle,** *University of Illinois at Urbana–Champaign (1976)*

2. *The Journal of Symbolic Logic 45 (3) p. 464–482. (Sep.,1980)*

3. *Mario Livio;* **The equation that could not be solved,** *Simon & Schuster* (Sep., 2006) (language of symmetry: group theory)

4. *Jhon A Mackan;* **Universe is Only Space Time,** *California, P 4 (2015)*

5. *Lee Smolin;* **Positive energy in quantum gravity,** *phys. rev d 90 no.4 p 1– 3 (2014)*

6. *Ashok Saxena;* **Our universe and how it works; (Quantum gravitation and Fifth dimension)** *2015 (India) hard copy (Manas Prakashan) 2017 Book baby e-book. CODATA 2018 values*

7. Michael Gillon, Amaury H.M.J. triaud, Dipier queloz, **Seven temperate terrestrial planets around nearby ultra cool dwarf star TRAPPIST 1,** 2017 *Nature* 542, 456–460.

8. *Ashok Saxena.* **Inside a Wave,** *2005, (India) Manas Prakashan.*

9. Paul Davies, **The new physics.** *Cambridge University Press.
    pp. 187–.ISBN 978-0-521-43831-5. Retrieved 1 May 2011*

10. *Stephen Hawking's << THE UNIVERSE IN A NUTSHELL >>
    ISBN-978 0593 048153, Transworld publishers. 2001*

RECOMMONDED READINGS

11. *Alan H. Guth.* **The inflationary universe: the quest for a new
    theory of cosmic origins.** *(1998) Basic Books. pp. 186–ISBN
    978-0-201-32840-0.*

12. *V.F. Mukhanov,* **Physical foundations of cosmology.** *(2005)
    Cambridge University Press. pp. 58–. ISBN 978 – 0-521-56398-7.*

13. Ryan Samaroo, **On Identifying Background-Structure in
    Classical Field Theories,** 2011, *Philosophy of Science,* 78
    (5):1070–1081.

14. Peter Fisher Epstein, **Shape Perception in a Relativistic
    Universe.**2018, *Mind* 127 (506):339–379.

15. *Lineweaver, Charles; Tamara M. Davis* **Misconceptions about the
    Big Bang.** *(2005). Scientific American. 292(3): 36 – 45. Bibcode:
    2005 SciAm.292c.36L.doi:10.1038/scientificamerican0305–36.*

16. *Krauss, Lawrence M.; Robert J. Scherrer.* **The Return of a Static
    Universe and the End of Cosmology.** *General Relativity and
    Gravitation.39(2007)(10):*

    1545–1550. arXiv:0704.0221.

17. *Paul Davies.* **The Goldilocks Enigma. First Mariner Books.** *p.
    43–.ISBN 978-0-618-59226-5.*

18. *Michio Kaku.* **Parallel Worlds: A Journey through Creation,
    Higher Dimensions, and the Future of the Cosmos.** *(2006)
    Knopf Doubleday Publishing Group. p. 385. ISBN 978-0-307-
    27698-8.*

19. *Bernard F. Schutz.* **Gravity from the ground up.** *(2003) Cambridge University Press. pp. 361– ISBN 978-0-521-45506-0.*

20. Brooke Alan Trisel, **How Human Life Matters in the Universe: A Reply to David Benatar.** 2019, *Journal of Philosophy of Life* 9 (1) 1–15.

21. Adolf Grünbaum, **The Pseudo-Problem of Creation in Physical Cosmology.** 1989, *Philosophy of Science* 56 (3):373 – 394.

22. Chris Smeenk, **False Vacuum: Early Universe Cosmology and the Development of Inflation.** 2005

23. Jean Eisenstaedt & A. J. Knox (eds.), **The Universe of General Relativity.** Boston: Birkhauser. pp. 223–257.

24. Abriel Vacariu & Mihai Vacariu, **Dark Matter and Dark Energy, Space and Time & Other Pseudo-Notions in Cosmology.** 201

25. Maya Lincoln, Avi Wasser, **Spontaneous Creation of the Universe Ex Nihilo.** 2014, *Physics of the Dark Universe* 2 (4):195–199.

26. Brian D. Josephson, **Limits to the Universality of Quantum Mechanics.** 1988, *Foundations of Physics* 18 (12):1195–1204.

27. Oleg Bazaluk, **World Existence and "Evolved Matter" as its Modern Model.** 2009, *Philosophy and Cosmology* 1 (1): 3–37

28. Smorodinsky Y.A., **TEMPERATURE,** 1988

29. *Itzhak Bars; John Terning,* **Extra Dimensions in Space and Time.** *(November 2009), Springer. pp. 27– ISBN 978-0-387-77637-8.*

30. *Robert P Kirshner.* **The Extravagant Universe: Exploding Stars, Dark Energy and the Accelerating Cosmos.** *(2002) Princeton University Press. p.71. ISBN 978-0-691-05862-7.*

31. *Carroll, Bradley W. Ostlie, Dale A.* **An Introduction to Modern Astrophysics** *(International ed.). (2013), Pearson. pp. 1173–1174. ISBN 9781292022932.*

32. *Planck Collaboration, "Planck 2015 results. XIII.* **Cosmological parameters** *(See Table 4 on page 32 of pdf)". (2016). Astronomy & Astrophysics. 594: A13. arXiv:1502.01589*

33. *Alfred Driessen,* **The Question of the Existence of God in the Book of Stephen Hawking: A Brief History of Time.** 1995, *Acta Philosophica* 4 (1): 83–93.

34. *Paul Davies,* **The Accidental Universe,** 1982, *Cambridge University Press* ISBN 0-521-24212-6

35. *Rinat M. Nugayev,* **Einstein's 1905 'Annus Mirabilis': Reconciliation of the Basic Research Traditions of Classical Physics.** 2019, *Axiomathes* 29 (3): 207–235.

36. *Władysław Krajewski,* **On the Interpretation of the Equation E = Mc²; Reply to Flores.** 2006, *International Studies in the Philosophy of Science* 20 (2): 215–216.

37. *Alexander Klimets,* **Philosophical Model of Special Relativity.** 2012 – *Quantum Magic* 9 (3): 3113–3123.

38. *Ryan Samaroo,* **There Is No Conspiracy of Inertia,** 2018, *British Journal for the Philosophy of Science* 69 (4): 957–982

39. *Loeb, Abraham.* **The Long-Term Future of Extragalactic Astronomy.** *Physical Review D. 65 (4). arXiv:astro-ph/0107568.*

40. *Kazanas, D.* **Dynamics of the universe and spontaneous symmetry breaking.** *(1980) The Astrophysical Journal. 241: L59 L61.*

41. *Bielewicz, P. Banday, A.J. Gorski, K.M. (2013). Auge, E. Dumarchez, J. Tran Thanh Van, J. (eds.).* **Constraints on the Topology of the Universe.** *Proceedings of the XLVIIth Rencontres de Moriond. 2012 (91). arXiv:1303.4004.*

42. *Abbott, Brian.* **Microwave (WMAP) All-Sky Survey.** *(2007) Hayden Planetarium, Retrieved.*

43. *E. Komatsu1et.al,* **Five-Year Wilkinson Microwave Anisotropy Probe Observations: Cosmological Interpretation,** 2009 February, *The Astrophysical Journal Supplement Series,* 180:330–376.

44. *Bennett, C.L. Larson, D. Weiland, J.L. Jarosik, N. et al. (1 October 2013).* **Nine-year Wilkinson Microwave Anisotropy Probe (WMAP) Observations: Final Maps and Results.** *The Astrophysical Journal Supplement Series. 208 (2)20. arXiv: 1212.5225.*

45. *Vaudrevange; Starkmanl; Cornish; Spergel* **Constraints on the Topology of the Universe: Extension to General Geometries.** *(2012). Physical Review D. 86 (8): 083526. arXiv:1206.293*

46. *Fixsen, D.J.* **The Temperature of the Cosmic Microwave Background.** *(2009) The Astrophysical Journal. 707(2): 916 – 920. arXiv:0911.1955.*

47. *Horvath, I. Hakkila, J. Bagoly, Z,* **The largest structure of the Universe, defined by Gamma-Ray Bursts.** *(2013). arXiv:1311.1104.*

48. *Jarrett, T.H.* **Large Scale Structure in the Local Universe: The 2MASS Galaxy Catalog.** *(2004) Publications of the Astronomical Society of Australia. 21 (4): 396–403.*

49. *Science.nasa.gov.***Impact on cosmology of the celestial anisotropy of the short gamma-ray bursts.** *(2009). Baltic Astronomy. 18: 293–296.*

50. *Gott, III, J.R. et al.* **A Map of the Universe.***(May 2005) The Astrophysical Journal. 624 (2): 463–484. arXiv:astro-ph/03105*

51. *Codata 2018 values*

Index

'Quantum gravitation theory' could only be constructed, with Planck's parameters, which are used in this model to construct the working model of universe as they include Gravitational constant 'G', Quantum constant 'h' and Speed of light 'c'.

PLANCK'S UNITS

Planck length $lp = \sqrt{hG/c^3} = 1.616199 \times 10^{-35}$ m

Planck mass $mp = \sqrt{hc/G} = 2.17651 \times 10^{-8}$ kg

Planck time $tp = \sqrt{hG/c^5} = 5.39106 \ 0.53896896 \times 10^{-44}$ s

Planck energy $Ep = \sqrt{hc^5/G} = 1.956 \times 10^9$ J

Planck angular frequency $hp = \sqrt{c^5/hG} = 1.855394405 \times 10^{43}$ s-1

Planck force $Ep/lp = \sqrt{c^4/G} = 1.21027 \times 10^{44}$ N

Planck energy density $Up = Ep/lp^3 = c^7/hG^2 = 4.636 \times 10^{113}$ J/m3

Planck temperature $Ep/kB = 1.417 \times 10^{32}$ °K

Number of Bound Treos in One Meter

1/Planck's least length i.e. number or bound treos per meter

One/1.615788303 $\times 10^{-35}$ meter = 0.618892956 $\times 10^{35}$ bound treos per meter

Number of Bound Treos in Faco Meter (10^{-15} meter)

$0.618892956 \times 10^{20}$ bound treos per faco meter

Number of Bound Treos in centimeter

$0.618892956 \times 10^{33}$ bound treos per Centimeter

Number of Bound Treos in one km.

$0.618892956 \times 10^{38}$ bound treos per Km

Number of Bound Treos in one Light Year

$60 \times 60 \times 24 \times 364.25 \times 1.855394405 \times 10^{43} = 5.8379 \times 10^{51}$ bound treo distance per year

Number of Bound Treos in one Mega Parse One parse is 3.26 light year

$3.26 \times 5.8379 \times 10^{51} = 19.031 \times 10^{51}$ bound treo distance is one parse = 19.031×10^{57} bound treo distance is one mega parse

Number of Bound Treos in Radius of Present Size of Universe 10^{62} bound treo distance radius = 13.7 billion light years

Number of Bound Treos radius which will be the Size of One Fully Expanded Future Universe before big crunch

S^2 or 10^{86} bound treo distance radius

Total life Span of Universe

$3.441937353 \times 10^{86}$ vibrations or $1.855394405 \times 10^{43}$ seconds or S seconds

Number of vibrations in one second (cosmic rhythm)

One second/Planck's least time = $1/0.53896896 \times 10^{-43}$ sec.

= $1.855394405 \times 10^{43}$ vibrations per second (Planck's frequency)

Birth of Photon

Each atom constantly absorbs energy and releases photon at its unique frequency; by jumping of an electron from higher to lower energy shell by one (or more) quantum level (s).

Energy of photon born = -2.18×10^{-18} ($1/n^2$ final $- 1/n^2$ initial)

its **Frequency = 4.6×10^{14} Hz**

its **Wave length = 6.56×10^{-7} m**

(Wave length = 6.56×10^{-7} m $\times$ 0.618892956 $\times 10^{35}$

Bound treos per meter = **4.0599378×10^{28} Bound treos)**

Speed of Light

Number of Vibrations in one Second

One second/Planck's least time = $1/0.53896896 \times 10^{-43}$ = 1.855394405 $\times 10^{43}$ vibrations per second (S vibrations)

in Planck's least time by one vibration the photon packet is pushed from 'one bound treo' to 'next bound treo' on space matrix.

then speed of photon = $1.855394405 \times 10^{43}$ i.e. S bound treo distance per second

Speed of Displacement of Photon Per Second

$1.855394405 \times 10^{43}$ vibrations per second $\times 1.615788303 \times 10^{-35}$ Meter[32] displacement per vibration = *2.997924577×10^8 meter per second, is conventional value of 'c'; the Speed of light.*

---

32 (1.615788303 $\times 10^{-35}$ meter (one Planck's distance) is occupied by one bound treo.)

Reduced Planck's Constant and Planck's Constant.

Reduced Planck constant h (h bar) is the value of, one quantum energy

1. **Number of Treos in one Quanta Energy**

   $1.855394405 \times 10^{43}$ treos (S number of free treos)

2. **Mass of Treos in one Quanta Energy**

   $1.855394405 \times 10^{43}$ free treos in one quanta/$1.581858906 \times 10^{94}$ number of treos in one Kg

   $1.172920289 \times 10^{-51}$ Kg

3. **Energy of one quanta mass ($e = mc^2$)**

   $1.172920289 \times 10^{-51}$ Kg $\times$ (2.997924577×10^8 meter per second)2

   $10.54168182 \times 10^{-34}$ Joule–Sec

   Planck Constant (h) is the Angular Momentum of one Quantum Energy

   10.54168182 Joule–Sec $\times 6.285714286$ (**value of 2π**)

   $= 6.6262 \times 10^{-34}$ Joule–Sec

(Exact conventional value of Planck's constant)

Gamma Photon and Pair Production

2.8724104×10^{64} free treos are present in 1.02 Mev energetic gamma photon of $1.54814006 \times 10^{21}$ frequency. It will divide in two equal parts of electron and positron pair, each having half ($1.439491604 \times 10^{64}$) free treos.

1. **Mass Energy of Gamma Photon of 1.02 Mev**

   $1.855394405 \times 10^{43}$ free treos per quantum $\times 1.54814006 \times 10^{21}$ frequency of wave = **2.8724104×10^{64} free treos.**

Or

0.181584489 × 10^{-29} kg calculated mass energy in gamma photon ×

1.581858906 × 10^{94} free treos per kg = **2.8724104 × 10^{64} free treos.**

2. **Energy of Gamma Photon of 1.02 Mev**

    0.181584489 − 10^{-29} kg × (2.997924577 × 10^8 meter per second)2 = **1.632 10^{-13} Joule−Sec (e = mc²)**

    or

    1.02 Mev × 1.6 × 10^{-19} J per electron = **1.632 10^{-13} Joule− Sec**

    Mass energy of Unit Electron of 0.51Mev

    1.581858906 × 10^{94} free treo per kg × 9.1 × 10^{-31} kg = **1.439491604 × 10^{64} free treos**

3. **Energy of Unit Electron of 0.51Mev**

    0.51 Mev × 1.6 × 10^{-19} J per electron = **0.816 × 10^{-13} joule**

Compton Wave Length of (0.51Mev) Electron

(2pr is circumference of circle or Compton wave length, while radius r is 'Reduced Compton wave length of electron')

2.4263102389(16) × 10^{12} m Compton wave length of Electron =

1.5016263 × 10^{23} bound treo distance

Reduced Compton Wave Length of (0.51Mev) Electron

(Rq quantum radius)

(Compton wave length/2 π)

(2.4263102389(16) × 10^{12}m/2 π = 0.38600391 × 10^{-12} M)

23.88951 × 10^{21}Bound treo distance

1. 'Reduced Compton wave length' of Electron × 137 (1/fine structure constant)

   = $5.29 \times 10^4 fm$ = 32.7225276 × 10^{23} Bound treo distance is **Bhor radius of Electron** (0.51Mev)

2. 'Reduced Compton wave length' of electron × 1/137 (Fine structure constant)

   2.82 fm =1.7437599 × 10^{20} **Bound treo distance is Classical radius of Electron** (0.51Mev)

Classical radius of Electron (0.51Mev)

   2.82 fm = 1.7437599 × 10^{20} **bound treo distance**

Bhor radius of bhor orbit in hydrogen atom

   5.29 ×10^4 fm = 32.7225276 × 10^{23} **bound treo distance**

Electron Charge

   1.6022 × 10^{-19} Coulomb (elementary charge or charge on one Electron)

Characteristics of an Electron

3.8616 ×	*10^{13} m*	*Rq quantum radius*
7.7634 ×	*10^{20} s-1*	*sc Compton angular frequency*
1.2356 ×	*10^{20} Hz*	*sc Compton frequency*
8.1871 ×	*10^{14} J*	*Ei internal energy*
9.1094 ×	*10^{31} kg*	*me electron's mass*
1.6022 ×	*10^{19} Coulomb*	*e elementary charge*
6.7635 × 10^{58} (Ei/Rq)		*Rs classical Schwarzschild radius*

Mass Energy of Proton

$1.672313038 \times 10^{-27}$ kg $\times 1.5808523 \times 10^{94}$ Free Treos per Kg = $2.6417043\ 73 \times 10^{67}$ free Treos

Proton = $2.641704371 \times 10^{67}$ free Treos

Mass Energy of Neutron

1.67×10^{-27} kg $\times 1.5808523 \times 10^{94}$ Free Treos per Kg

$= 2.645363273 \times 10^{67}$ free Treos

Neutron = $2.645363273 \times 10^{67}$ free Treos

Mass Energy of W Boson

$14.24192075 \times 10^{-26}$ kg $\times 1.5808523 \times 10^{94}$ Free Treos per Kg

$= 22.52870912 \times 10^{68}$ free treos

$22.52870912 \times 10^{68}$ free treos or 80 Gev

Mass Energy of Z Boson

$16.02216084 \times 10^{-26}$ kg $\times 1.5808523 \times 10^{94}$ Free Treos per Kg

$= 25.34479782 \times 10^{68}$ free treos

$25.34479782 \times 10^{68}$ free treos or 90 Gev

Mass Energy of Unit Mass (Planck Mass)

One–unit mass is = **$2.17651\ 10^{-8}$ kg $\times 1.5808523 \times 10^{94}$ free treos per kg = $3.442488398 \times 10^{86}$ free treos, i.e. S^2 free treos.**

Mass Energy of Carbon Atom

$2.641704373 \times 10^{67} \times 12$ nucleons = 31.700×10^{67} free treos

Mass of one kg Carbon

5.0188×10^{25} number of carbon atom in one kg × 31.700×10^{67} Free Treos in one carbon atom.

1.5808523×10^{94} Free Treos

Mass Energy in one kg weight

One Kg mass = 1. 5808523×10^{94} Free Treos

Mass Energy in one Unit Black Hole

S^3 free treos or S unit masses

$(1.855394405 \times 10^{43})^3$

$6.3871737 \times 10^{129}$ free treos

Distance of Planetary orbit of Earth from Sun (semi major axis of Earth)

$0.9258638622 \times 10^{46}$ bound treo distance from Sun Speed of Revolution of Earth

29.7934848 km per second = $1.84389778 \times 10^{39}$ bound treo distance per sec

Size of Planetary Orbit of Earth

Circumference of orbit = 2r × π

[2 × $0.9258638622 \times 10^{46}$ −1 bound treos × π]

= 5.8197156×10^{46} bound treos form orbit of planet Earth

Or

Orbital speed of earth $1.84389778 \times 10^{39}$ bound treo distance per second × 31557600 seconds in one year = 5.8188989×10^{46} bound treo distance in one year.

ELOBORATING FINE STRUCTURE CONSTANT

(1) Reverse fine structure constant ≈ i.e. 137 is 33 ᵗʰ prime number.

(2) Reverse fine structure constant ≈ 0.1 × 1.11×11.111 × 111.1111 = 137.035

(3) 137 prevails in universe because it makes golden ration (1: 1.6803)

if you make 137 degree arc in a circle and **arc in rest of circumference of a circle,** *two are in* **golden ratio.** Golden ratio 1:1.62 prevails in universe and emply in carving of different shapes of flora (1, 3, 5, 7 pattern of florosence) and fauna, governs musical notes, structure of pyramids, shape of egg and even our egg shaped universe.

(4) value of Fine structure constant is calculated = Ke²/hc = coulomb × Square of charge/speed of light × Planck constant = 7.29735254 ×10⁻³= 1/137

(5) **Single electron around hydrogen atom moves in bhor orbit moves at, speed of light/137,** *at about 2200 km per second* **and in any element** *with its atomic number Z electron moves* **Z/137 times the speed of light.**

(6) If you **divide RC wave length of electron by fine structure constant** *you calculates the* **radius of Bhor orbit**

(7) If you **multiply RC wave length of electron by fine structure constant** *you calculates* **classical radius of electron.**

(8) Energy of hydrogen atom is E1= e²/ 2a (where e is charge and a is fine structure constant i.e. 1/137). When it is converted in jule it is **value of Rydberg constant RH.**

(9) It is used to calculate Rydberg constant R ω = **cube of fine structure constant divided by 4 π r.**

(11) If you divide energy of hydrozen atom by fine structure constant you calculates mass of hydrogen atom.

(12) Electron energy 0.51 MeV divided by twice of fine structure constant calculates the composite mass energy of electron C mue (mass energy + localisation energy i.e. **35.01MeV**) and **integral multiple of** this basic **Cmue units** form all matter (elementary particles and nucleons)

(13) Atomic structures are produced in a pyramid shaped area produced by over lapping of increasing deformation and counter deformation in second dimension, **in golden ratio pyramid shaped deformation** (in which 'area of base of pyramid' and its 'height of pyramid' is in golden ratio) at four quantum levels (total 8 quantum levels of deformation and counter deformation) forms all 118 elements of periodic tabel.

www.ingramcontent.com/pod-product-compliance
Lightning Source LLC
Chambersburg PA
CBHW021352210526
45463CB00001B/72